Lecture Notes
in Control and Information Sciences

213

Editor: M. Thoma

Springer-Verlag London Ltd.

Amit Patra and Ganti Prasada Rao

General Hybrid Orthogonal Functions and their Applications in Systems and Control

 Springer

Authors

Ganti Prasada Rao, Professor, Dr
Amit Patra, Dr
Department of Electrical Engineering, Indian Institute of Technology
Kharagpur 721302, India

ISBN 978-3-540-76039-9

British Library Cataloguing in Publication Data
Patra, Amit
   General hybrid orthogonal functions and their applications
   in systems and control. - (Lecture notes in control and
   information sciences ; 213)
   1.Functions, Orthogonal 2.Orthogonal polynomials
   3.Mathematical analysis
   I.Title II. Rao, Ganti Prasada
   515.5'5
   ISBN 978-3-540-76039-9   ISBN 978-3-540-40950-2 (eBook)
   DOI 10.1007/978-3-540-40950-2
Library of Congress Cataloging-in-Publication Data
A catalog record for this book is available from the Library of Congress

Typesetting: Camera ready by author

69/3830-543210 Printed on acid-free paper

To our families,
with love and affection

Biodata of the authors

Amit Patra was born in Kharagpur, India on October 1, 1962 and received the B.Tech., M.Tech. and Ph.D. degrees in Electrical Engineering from Indian Institute of Technology, Kharagpur, in 1984, 1985 and 1990 respectively.

In 1987 he joined the Department of Electrical Engineering, Indian Institute of Technology, Kharagpur as a Lecturer and became an Assistant Professor in 1990. From October, 1992 to December, 1993, he visited the Lehrstuhl für Elektrische Steuerung und Regelung, Ruhr Universität Bochum, Germany, as a Research Fellow of the Alexander von Humboldt Foundation. He has published more than 40 research papers in various International Journals and National and International Conferences. His fields of interest are system identification, industrial automation and control and discrete-event systems.

He received the I.I.T. Kharagpur Silver Medal and I.I.T. Kharagpur Technology Alumni Association Gold Medal for academic excellence in his undergraduate and post-graduate studies respectively. He has received the Young Teachers' Career Award from the All India Council of Technical Education.

Dr. Patra is a member of IEEE(USA), IE(India) and a life-member of Systems Society of India. He has also been selected as an Associate of the Indian Academy of Sciences for the period 1992-1997.

Ganti Prasada Rao was born in Seethanagaram, Andhra Pradesh, India, on August 25, 1942. He studied at the College of Engineering, Kakinada, and received the B.E. degree in Electrical Engineering from Andhra University Waltair, India, in 1963, with first class and high honours. He received the M.Tech. (Control Systems Engineering) and Ph.D. degrees in Electrical Engineering in 1965 and 1969 respectively, both from the Indian Institute of Technology, Kharagpur, India.

From July 1969 to October 1971 he was with the Department of Electrical Engineering, PSG College of Technology, Coimbatore, India, as an Assistant Professor. In October 1971 he joined the Department of Electrical Engineering, Indian Institute of Technology (IIT) Kharagpur, as an Assistant Professor and became a Professor in May 1978. From May 1978 to August 1980, he was the Chairman of the Curriculum Development Cell (Electrical Engineering) established by the Government of India at IIT Kharagpur. From October 1975 to July 1976, he was with the Control Systems Centre, University of Manchester Institute of Science and Technology (UMIST), Manchester, England, as a Commonwealth Postdoctoral Research Fellow. From October 1982 to November 1983, and again during May-June 1985, and May-July 1991, he was with the Lehrstuhl für Elektrische Steuerung und Regelung, Ruhr-Universität Bochum, Germany, as a Research Fellow of the Alexander von Humboldt Foundation. Since June 1992 he is with the Engineering Systems Division, Power and Desalination Plants, Water and Electricity Department, Government of Abu Dhabi, United Arab Emirates as a Scientific Advisor. He has research interests and publications in the areas of mathematical instruments, time varying systems, parametric phenomena, system identification, fuzzy logic control, piecewise constant and generalised hybrid orthogonal functions, large scale systems, continuous time approaches to system identification and adaptive control. He has authored three books: *Piecewise Constant Orthogonal Functions and Their Applications to Systems and Control, Identification of Continuous Dynamical Systems - The Poisson Moment Functional Approach* (with D.C.Saha) both published by Springer Verlag in 1983, and *Identification of Continuous Systems* (with H.Unbehauen), published by North Holland, in 1987. He is co-editor (with N. K. Sinha) of *Identification of Continuous-Time Systems - Methodology and Computer Implementation*, Kluwer, 1991.

He received several academic awards including the IIT Kharagpur Silver Jubilee Research Award of 1985.

Professor Rao is a Member, Editorial Board, *IASTED International Journal of Modelling and Simulation, Systems Science*, Journal of the Polish Academy of Sciences and *IETE (India) Students' Journal*. He is also Member, Honorary Education and Research Advisory Boards, American Biographical Institute (ABI). He was Chairman, Technical Committee, 13th National Systems Conference, 13-15 December, 1989 held at I.I.T. Kharagpur. He is associated with several National and International Conferences as a member of their Steering and Program Committees. He was guest editor (with H. Unbehauen) of the special issue on *Identification and Adaptive Control - Continuous-time Approaches* of Control Theory and Advanced Technology, March 1993. At several IFAC Symposia he organized (with Prof. H. Unbehauen) many invited sessions on continuous-time approaches to system identification. He was also guest-editor, Special issues of IETE (India) Students' Journal, Part I, October 1992 and Part II, January-March 1993.

Professor Rao is a Life Fellow of IE (India), a Fellow of the IETE (India), Senior Member of the IEEE (USA) and a Fellow of the Indian National Academy of Engineering.

Contents

Preface   iii

List of Symbols   vii

List of Abbreviations   xii

1 Introduction   1

   1.1 State of the art 1

   1.2 Definition of the system of general hybrid orthogonal functions 3

   1.3 Properties of GHOF 5

      1.3.1 Orthogonality 5

      1.3.2 Function expansion 6

      1.3.3 Completeness 6

2 GHOF Spectral Analysis of Dynamical Systems   11

   2.1 Survey of literature in the field 11

   2.2 GHOF operational matrix for integration 13

   2.3 Solution of state equation 16

   2.4 Extension of solution beyond the initial interval 18

      2.4.1 Multiple Segment Multiple Term (MSMT) Formula 18

      2.4.2 Single Segment Multiple Term (SSMT) Formula . 19

      2.4.3 Multiple Segment Single Term (MSST) Formula . 20

2.4.4   Single Segment Single Term (SSST) Formula ... 20

2.5   General framework of numerical analysis of dynamical systems 21

2.6   Simulation of SCR-controlled DC drives 30

2.7   Prediction of limit cycle of van der Pol's oscillator 41

2.8   Remarks 46

3   Identification of Continuous-time Systems   47

3.1   Survey of literature in the field 47

3.2   Formulation of the problem in terms of GHOF spectra . . 50

3.3   Recursive computation of multiple integrals of a signal . . 52

3.4   Recursive least squares (LS) parameter estimation algorithm employing GHOF 54

3.5   Parameter estimation in a converter driven DC motor system 55

3.6   Parameter estimation using generalized least squares (GLS) scheme 58

3.7   Simultaneous state and parameter estimation of SISO systems 63

3.8   Remarks 68

4   Continuous-time Model-based Self-tuning Control   71

4.1   Survey of literature in the field 71

4.2   The STC problem in a CT setting 74

4.3   Implementation of CT model-based STC 77

4.4   Remarks 80

5   Other Possible Applications   85

Bibliography   88

Index   117

Preface

The use of complete systems of orthogonal functions as bases of expansion for square integrable real-valued functions is a standard method in mathematical analysis and computational techniques. Several sets of orthogonal basis functions are available in mathematics and their applications are too numerous to be cited here. The existing sets of orthogonal functions can be broadly divided into two classes. One includes the classical sets of continuous functions and the other consists of piecewise constant systems having inherent discontinuities.

Sets of orthogonal polynomials (e.g., Legendre, Laguerre, Chebyshev, Jacobi, Hermite etc.) along with the well-known set of sine-cosine functions extensively used in the classical literature, are continuous over their intervals of definition and consequently are well-suited to approximate continuous functions. Piecewise constant systems of Walsh, block pulse and Haar functions are relatively more recent. They give rise invariably to staircase approximations of functions, introducing discontinuities according to the nature of the chosen basis.

The techniques of reducing the calculus of continuous dynamical systems to an attractive algebra, approximate in the sense of least squares and convenient for analysis and computation, have emerged in the early seventies, first mainly with reference to piecewise constant basis functions (PCBF), yielding solutions to several problems of systems and control. The developments were comprehensively covered in a book by G. P. Rao in 1983 [216]. In the subsequent years the use of continuous basis functions (CBF) in similar problems and situations has been amply demonstrated in several publications.

In the meantime, investigations into the effectiveness of basis functions in expanding certain functions of the real world, such as those arising in the treatment of power electronic circuits and systems, revealed the inadequacy of the PCBF and CBF each taken alone in meet-

ing the needs of reality, i.e., to match the mixed features of continuity
and jumps simultaneously. As a result of these investigations, the general
hybrid orthogonal functions (GHOF) have been recently proposed
by the authors. They form a very general and flexible framework of
orthogonal functions capable of modeling the mixed features of continuity
and jumps in functions encountered in certain important practical
situations. All the well-known sets of orthogonal functions in CBF and
PCBF can be derived as special cases or as linear transformations of
GHOF.

This monograph introduces the GHOF and illustrates their use as a
flexible framework of computational tools in a variety of relevant problems
in systems and control and is thus expected to be a naturally desirable
supplement to the book on PCBF [216]. An important feature of the
book is its coverage of recursive algorithms and a completely continuous-
time based self-tuning control scheme using block pulse functions which
are seen to belong to the GHOF family.

The authors are grateful to many of their colleagues at home and
abroad for their significant contributions which were instrumental in
shaping the ideas presented in this work. In particular, they would
like to express their gratitude to Professor H. Unbehauen, Lehrstuhl
für Elektrische Steuerung und Regelung, Ruhr University, Bochum, for
providing the facilities required at the final stage in the preparation
of the monograph. They thank the authorities of the Indian Institute
of Technology, Kharagpur for the facilities and the right atmosphere
provided for research. The authors are indebted to their families for
their patience, understanding and encouragement.

January, 1995   A. Patra
Kharagpur   G. P. Rao

List of Symbols

$(\cdot, \cdot)$	Inner product of two functions / an interval
$\{\cdots\}$	Set consisting of $\cdots$
$[a_i]$	Matrix or vector comprising a_i, $\forall i$
$\otimes$	Kronecker product of two matrices
$\forall j$	For all j
$\cup_j$	Union over all j
$\mathbf{A}^T$, $\mathbf{a}^T$	Transpose of a matrix $\mathbf{A}$ or vector $\mathbf{a}$
$\mathbf{A}_{(k \times l)}$	Matrix $\mathbf{A}$ of dimension $k \times l$
$\hat{A}, \hat{a}$	Estimate of A or a
a_i	Coefficient of $\dfrac{d^i y}{dt^i}$
A, $A(s)$	System denominator polynomial in s
A_m, $A_m(s)$	Model denominator polynomial in s
$\mathbf{A}$	System matrix in state space realizations
$\mathcal{A}$	Approximation of $\mathcal{I}$
b_i	Coefficient of $\dfrac{d^i u}{dt^i}$
B, $B(s)$	System numerator polynomial in s
$\mathbf{B}$	Input matrix in state space realizations
$\mathcal{B}$	Boundary condition
c_i	Coefficient of s^i in $C(s)$
C, $C(s)$	Noise polynomial in s
$\mathbf{C}$	Output matrix in state space realizations
d_i	Coefficient of s^i in $D(s)$
D, $D(s)$	Disturbance polynomial in s
$\mathcal{D}$	Differential equation

$e(t)$, $e(k)$	Equation error
$\mathbf{e}$	Error vector
$\mathbf{E}$	Operational matrix for integration for Legendre Poly
$\mathbf{E}_j$	Operational matrix for integration for CBF
$\mathbf{E}_b$	Operational matrix for integration for BPF
$\mathbf{E}_g$	Operational matrix for integration for GHOF
$\mathcal{E}_f$	Integral of squared error corresponding to $f(t)$
$\mathcal{E}_y$	Normalized integral of squared error corresponding to $y(t)$
f, $f(t)$	Function of t
$f_{i,j}$	GHOF spectral coefficients of $f(t)$
$\mathbf{f}$	Vector comprising $f_{i,j}$, $\forall i, j$
$\mathbf{f}_j$	Vector comprising $f_{i,j}$, $\forall i$ and a given j
F, $F(s)$	Feedback polynomial
g, $g(t)$	Function of t, (sometimes) integral of $f(t)$
$g_{i,j}$	GHOF spectral coefficients of $g(t)$
$\mathbf{g}$	Vector comprising $g_{i,j}$, $\forall i, j$
$\mathbf{g}_j$	Vector comprising $g_{i,j}$, $\forall i$ and a given j
G, $G(s)$	Controller polynomial
$\mathcal{G}$	The set comprising $\theta_{i,j}$, $\forall i, j$
$\mathcal{G}_j$	The set comprising $\theta_{i,j}$, $\forall i$ and a given j
h, $h(t)$	Integral of $g(t)$
$h_{l,j,k}$	Elements of $\mathbf{H}_{l,j}$
$\mathbf{h}$	GHOF spectral vector corresponding to $h(t)$
H, $H(s)$	Controller (detuning) polynomial
$\mathbf{H}_{l,j}$	Elemental blocks of $\mathbf{E}_g$
$i(t)$	DC motor armature current
$\mathbf{I}$	Identity matrix of appropriate size
$\mathcal{I}$	Integral equation
$\mathcal{I}_n$	The set $\{1, 2, \ldots, n\}$ where n is an integer
$\mathcal{I}_n^k$	The set $\{k, k+1, \ldots, n\}$ where k and n are integers
J	Moment of inertia of a DC motor system
$\mathcal{J}$	Functional to be minimized
k_n	The ratio of back e.m.f. to line voltage in a DC motor
K_a, K_T	Armature and torque constants of a DC motor

L	Armature inductance of a DC motor
$\mathbf{L}_{2\rho}$	Space of ρ-weighted square integrable functions
m	Number of segments in GHOF
m_r	Covariance matrix resetting interval
$M, M(s)$	Model transfer functions
$\mathbf{M}$	Measurement matrix
n	Order of a system
n_b	Order of the numerator of a transfer function
n_d	Order of the polynomial $D(s)$
n_i	Number of inputs of a system
n_o	Number of outputs of a system
$\mathbf{N}(\cdot)$	Nonlinear operator
$p_{i,j}(t)$	Elements of CBF
$\mathbf{p}_j$	Vector of $p_{i,j}$, $\forall i$ and a given j
$\mathbf{p}, \mathbf{p}(t), \mathbf{p}(k)$	Parameter vector
$\mathbf{P}, \mathbf{P}(k)$	Covariance matrix
q	Forward shift operator
$q_{i,j}$	Inner product of $\theta_{i,j}$ with itself
$\mathbf{q}$	Kalman gain vector
$Q, Q(s)$	Polynomial in s
$r, r(t)$	Integral of $f^3(t)$
r_j	Number of CBF components over the j-th segment
$\mathbf{r}$	GHOF spectral vector corresponding to $r(t)$
R	Armature resistance of a DC motor
$\mathbf{R}$	Matrix element of $\mathbf{E}_g$
$\mathbf{R}_1$	One-dimensional real Euclidean space
s_{ijk}	GHOF spectral coefficients of $\mathbf{B}u(t)$
$\mathbf{s}_{ij}$	Vector comprising s_{ijk}, $\forall k$
$S, S(s)$	Feedforward polynomial
$\mathbf{S}$	Matrix comprising $\mathbf{s}_{i,j}$, $\forall i, j$
$\hat{\mathbf{S}}$	Matrix defined in (2.20)
$\mathcal{S}$	Vector constructed with the column of $\hat{\mathbf{S}}$

t	Real variable (normally corresponds to time)
T	Interval of definition of GHOF
T_i	Widths of GHOF segments
T_L	Load torque
$u, u(t)$	Input of a SISO system
$u_i(t)$	The i-th element of $\mathbf{u}(t)$
$u^{(i)}(t)$	Shorthand for $\dfrac{d^i u}{dt^i}$
$u_{(i)}(t)$	The i-th integral of $u(t)$
$\mathbf{u}$	GHOF spectral vector for $u(t)$
$\mathbf{u}(t)$	Input vector of a MIMO system
$U, U(s)$	Laplace transform of $u(t)$
$v, v(t)$	Disturbance signal / Equation error in GLS scheme
v_{ijk}	GHOF spectral coefficients of $\dot{\mathbf{x}}(t)$
$\mathbf{v}_{ij}$	Vector comprising $v_{ijk}, \forall k$
$v_L(t)$	Line voltage applied to a DC motor
$v_t(t)$	Terminal voltage across a DC motor
$V, V(s)$	Laplace transform of $v(t)$
$\mathbf{V}$	Matrix comprising the GHOF spectra of $\dot{\mathbf{x}}(t)$
$\mathcal{V}$	Vector constructed with the columns of $\mathbf{V}$
$w, w(t)$	Set-point signal
$W, W(s)$	Laplace transform of $w(t)$
$x_i(t)$	The i-th element of $\mathbf{x}(t)$
$x_i^{(j)}(t)$	The j-th derivative of $x_i(t)$
x_{ijk}	GHOF spectral coefficients of $\mathbf{x}(t)$
$\mathbf{x}_{ij}$	Vector comprising $x_{ijk}, \forall k$
$\mathbf{x}, \mathbf{x}(t)$	State vector
$\mathbf{x}_0$	GHOF spectra of $\mathbf{x}(0)$
$\mathbf{X}$	Matrix comprising $\mathbf{x}_{i,j} \forall i, j$
$y, y(t)$	Output of a SISO system
$y_i(t)$	The i-th element of $\mathbf{y}(t)$
$y^{(i)}(t)$	Shorthand for $\dfrac{d^i y}{dt^i}$
$y_{(i)}(t)$	The i-th integral of $y(t)$

$\mathbf{y}$	GHOF spectral vector for $y(t)$
$\mathbf{y}(t)$	Vector of outputs in a MIMO system
$Y, Y(s)$	Laplace transform of $y(t)$
$\mathbf{Y}$	Matrix comprising the GHOF spectra of $\mathbf{y}(t)$
$z, z(t)$	Noise signal
$Z, Z(s)$	Laplace transform of $z(t)$
α	Firing angle of a converter
β	Extinction angle of a converter
γ_i	Initial condition terms in parameter estimation
ϵ	Coefficient in van der Pol's equation
ζ_j	Term defined in (3.16)
η_i	Approximation error term
η	Vector of η_i, $\forall i$
θ	Phase angle of an R-L circuit
$\theta_{i,j}$	Element of the set $\mathcal{G}$
$\iota, \iota(t)$	Unit step function
$\iota_{(i)}(t)$	The i-th integral of $\iota(t)$
ι	GHOF spectral vector for $\iota(t)$
κ	Vector defined in (2.27)
λ	$A_m(0)/B(0)$
μ	Small constant
ν	Vector defined in (2.25)
ν_k	Vector defined in (2.26)
ξ	Vector of unknowns
$\rho, \rho(k)$	Weighting/ Forgetting factor
$\rho(t)$	Weighting function
ϱ	Total number of elements in $\mathcal{G}$
σ	Large positive constant
$\phi_{i,j}, \phi(t), \psi$	Measurement vectors
$\omega, \omega(t)$	DC motor speed
Γ, Γ_j	Boundaries of domains Ω, Ω_j
Δ	Matrix defined in (2.11)
Θ	Vector comprising $\theta_{i,j}, \forall i, j$
Φ_j	Measurement matrix
Ω	Domain of definition of GHOF
Ω_j	The j-th segment (element) of Ω
Ψ, Ψ'	Vectors of weighting functions

List of Abbreviations

AC	Alternating Current
BPF	Block Pulse Functions
CBF	Continuous Basis Functions
CHP	Chebyshev Polynomials
CT	Continuous-time
DC	Direct Current
DT	Discrete-time
e.m.f.	Electromotive Force
FE	Finite Element
FWD	Free Wheeling Diode
GBPF	Generalized Block Pulse Functions
GHOF	General Hybrid Orthogonal Functions
GLS	Generalized Least Squares
GOP	Generalized Orthogonal Polynomials
HEP	Hermite Polynomials
HF	Haar Functions
JAP	Jacobi Polynomials
LAP	Laguerre Polynomials
LEP	Legendre Polynomials
l.h.s.	Left Hand Side (of an equation)
LQG	Linear Quadratic Gaussian
LS	Least Squares
LT	Linear Time-invariant
MF	Modulating Functions
MIMO	Multiple Input Multiple Output
MRAC	Model Reference Adaptive Control

MSMT	Multiple Segment Multiple Term
MSST	Multiple Segment Single Term
MV	Multivariable (System)
NI	Numerical Integration
NL	Nonlinear (System)
OF	Orthogonal Functions
OP	Orthogonal Polynomials
PCBF	Piecewise Constant Basis Functions
PID	Proportional-Integral-Derivative
PMF	Poisson Moment Functionals
PRBS	Pseudo Random Binary Sequence
r.h.s.	Right Hand Side (of an equation)
r.m.s.	Root Mean Square
SC	Scaled (System)
SCF	Sine Cosine Functions
SCR	Silicon Controlled Rectifier
SG	Singular (System)
SISO	Single Input Single Output
SNR	Signal to Noise Ratio
SS	Stiff (System)
SSMT	Single Segment Multiple Term
SSST	Single Segment Single Term
TD	Time Delay
TS	Taylor Series
TV	Time-Varying (System)
WF	Walsh Functions

Chapter 1

Introduction

1.1 State of the art

In the field of dynamic systems and control, orthogonal functions (OF)-based techniques of analysis, identification and control have received considerable attention in the recent years. This is evident from the vast amount of literature published over the last two decades. The use of sine and cosine functions (SCF) in Fourier (harmonic) analysis of applied mathematics is well-known for a long time but their application to typical problems of systems and control followed only after the success of various other systems of orthogonal functions in solving such problems.

The various systems of orthogonal functions may be classified into two categories. The first is the so-called piecewise constant basis functions (PCBF) to which the orthogonal systems of Walsh functions (WF), block pulse functions (BPF) and Haar functions (HF) belong. These functions are constant over different segments within their intervals of definition and the functions and solutions represented using this class as basis are always staircase-approximated. Despite this, these were extensively applied to many areas of systems and control [216,233] in the last decade and seem to have inspired the use of various systems of orthogonal polynomials in the recent past [197]. Orthogonal polynomials and the sine-cosine functions may be combined into a broader class, viz., that of continuous basis functions (CBF). The various systems of orthogonal polynomials such as Legendre, Laguerre, Chebyshev (of kind I and II), Jacobi, Hermite etc. along with sine-cosine functions belong to this class. These functions also have been applied with considerable success in the present decade. The problems considered so far for OF-based solutions include response analysis, optimal control, parameter estimation, model reduction, controller design, state estimation etc. They have been applied to linear time-invariant and time-varying systems, nonlinear and distributed parameter systems which include scaled systems, stiff systems, delay systems, singular systems and multivariable systems.

Because of the inherent discontinuous property of the systems of PCBF, they are efficient in representing discontinuous functions [177]. On the other hand, when the signals to be represented are actually continuous in nature, this basis will always give a stair-case fit, and to retain sufficient accuracy, a large number of terms will have to be retained in their expansions. In such situations, obviously the CBF will be a better basis. Therefore, in a given situation, one has to choose a set of basis functions depending on their suitability.

However, there are situations when none of these bases taken alone is adequate for efficient approximation of signals. For example, the types of signals encountered in the field of power electronics have mixed features of continuity and jumps. These signals (such as outputs of fully controlled bridge converters) are continuous over certain segments of time, with jumps occurring at the transitions of the segments. To meet these situations, one needs a suitable *hybrid* system of basis functions inherently possessing the required features of continuity mixed with jumps. The aim of this chapter is to present such a basis which is *general* enough to encompass both the classes of orthogonal functions and is *hybrid* in the sense that both the continuity and discontinuity properties can be modelled simultaneously.

One more limitation of most of the existing orthogonal-functions-based techniques is that the computational schemes for analysis, parameter estimation etc. are not suitable for real-time implementations. The spectra of the signals have to be computed corresponding to the entire interval of definition of the basis functions. These are therefore available only after this interval elapses, which is often a very large period of time. For real-time applications it is necessary to derive a recursive relation involving the spectra corresponding to successive segments of time within the overall interval. Among the existing classes of orthogonal functions, only block pulse functions possess this feature and have therefore been applied to some real-time problems. The basis proposed here also lends itself to such time-recursive formulations.

In the next section, a system of general hybrid orthogonal functions (GHOF) is defined. The subsequent sections present various properties of GHOF such as completeness, orthogonality and the formal method for function expansion. An illustrative example is given to compare the relative merits of PCBF, CBF and the proposed system of GHOF.

1.2 Definition of the system of general hybrid orthogonal functions

Consider the domain $\Omega = (0, T) \subset \mathbf{R}_1$ on which a set of general hybrid orthogonal functions (GHOF) is defined as

$$
\left.
\begin{aligned}
\mathcal{G} &= \{\theta_{i,j}(t) \mid j \in \mathcal{I}_m, i \in \mathcal{I}_{r_j}\}, \forall t \in \Omega \subset \mathbf{R}_1 \\
\text{where,} & \\
\theta_{i,j}(t) &= \begin{cases} p_{i,j}((t - t_{j-1})/T_j), & t \in \Omega_j = (t_{j-1}, t_j) \\ 0, & \text{otherwise} \end{cases}
\end{aligned}
\right\}, \quad (1.1)
$$

with $t_j = \sum_{l=1}^{j} T_l$, such that $T_j = t_j - t_{j-1}, \bigcup_{j=1}^{m} \Omega_j = \Omega, \mathcal{I}_n = \{1, 2, \ldots, n\}$ and the set $\mathcal{G}_j = \{\theta_{i,j}(t) \mid i \in \mathcal{I}_{r_j}\} = \{p_{i,j}((t - t_{j-1})/T_j) \mid i \in \mathcal{I}_{r_j}\}, \forall j \in \mathcal{I}_m$ is one of complete and orthogonal CBF defined over the segment Ω_j. There is complete flexibility in the choice of CBF in different segments Ω_j of Ω. That is, for a given j, $\mathcal{G}_j$ could be the system of Legendre polynomials, while for another j, one might choose the set of sine-cosine functions, if so desired. The total number of elements in the above set of GHOF is,

$$
\varrho = \sum_{j=1}^{m} r_j. \quad (1.2)
$$

From the above general definition, special cases may be seen to give the existing classes of CBF and PCBF as follows:

a) *Special case 1*

If $m = 1$,

$$
\mathcal{G} = \mathcal{G}_1 = \{p_{i,1}(t/T) \mid i \in \mathcal{I}_{r_1}\},
$$

which corresponds to a set of CBF defined over $(0, T)$. Orthogonal polynomials such as Legendre, Chebyshev and Jacobi, which are defined over

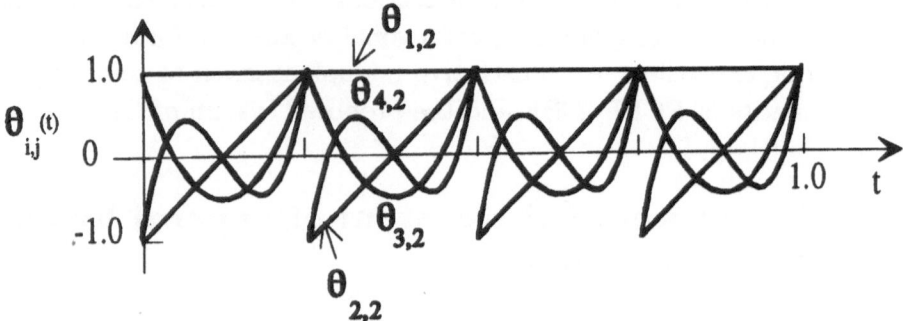

Figure 1.1: A set of GHOF comprising Legendre polynomials

a finite time-interval can be easily time-shifted and scaled to correspond to this normalized interval [87]. Sine-cosine functions, which are normally defined over the interval $(0, 2\pi)$ can also be appropriately scaled and used here. However, Laguerre and Hermite polynomials are defined over an infinite time-interval, and therefore they may be used only in a single segment framework.

b) *Special case 2*

If $r_j = 1, \forall j \in \mathcal{I}_m$,

$$\mathcal{G} = \{\theta_{1,j}(t) \mid j \in \mathcal{I}_m\}.$$

This is a set of generalized block pulse functions (GBPF) which belongs to the class of PCBF. Further, if

$$T_j = T/m, \forall j \in \mathcal{I}_m,$$

then $\mathcal{G}$ reduces to the well-known standard set of block pulse functions (BPF). It may be noted that the WF and HF [216] do not directly come under the GHOF framework. However, they are related to BPF by linear (orthogonal) transformations.

Fig. 1.1 shows, for instance, a set of GHOF with $m = 4, T = 1, r_j = 4, T_j = 0.25, \forall j \in \mathcal{I}_4$, in which $\mathcal{G}_j, \forall j$ is the set of Legendre polynomials of order upto 3.

1.3 Properties of GHOF

In this section certain essential properties of the system of GHOF, namely orthogonality and completeness and some aspects of function expansion are discussed.

1.3.1 Orthogonality

Let f and g be two functions belonging to the $\mathbf{L}_{2\rho}$ space defined over the domain Ω. Then the inner product between f and g is given by,

$$(f, g) = \int_{\Omega} \rho(t) \, f(t) \, g(t) \, dt \, , \qquad (1.3)$$

where $\rho(t)$ is the weighting function.

For $f(t)$ and $g(t)$ to be mutually orthogonal, their inner product (f, g) must be 0. To prove the orthogonality of the elements of $\mathcal{G}$ we must show that

$$(\theta_{i,j}(t), \theta_{k,l}(t)) = 0, \forall \, i, j, k, l \text{ such that } i \neq k \text{ or } j \neq l. \qquad (1.4)$$

The elements of $\mathcal{G}_j$ are orthogonal in pairs since the systems of CBF are orthogonal. That is,

$$(\theta_{i,j}(t), \theta_{k,j}(t)) = 0, \forall \, i, j, k \text{ such that } i \neq k. \qquad (1.5)$$

Again the sets $\mathcal{G}_j$ and $\mathcal{G}_l$ are disjoint in time, for $j \neq l$. Therefore,

$$(\theta_{i,j}(t), \theta_{k,l}(t)) = 0, \forall \, i, j, k, l \text{ such that } j \neq l. \qquad (1.6)$$

Equations (1.5) and (1.6) together imply (1.4).

However, in general, the system is not orthonormal, and

$$(\theta_{i,j}(t), \theta_{i,j}(t)) = q_{i,j}. \qquad (1.7)$$

The value of the constant depends on the chosen systems of CBF over each segment Ω_j. Each $\mathcal{G}_j$ is normalized (dividing $\theta_{i,j}$ by $q_{i,j}, \forall i, j$) for convenience in the following.

1.3.2 Function expansion

A function $f(t)$, belonging to the space $\mathbf{L}_{2\rho}$ defined over Ω, can be formally expanded as

$$f(t) = \sum_{j=1}^{m} \sum_{i=1}^{\infty} f_{i,j} \; \theta_{i,j}(t). \tag{1.8}$$

If $\mathcal{G}$ is orthonormal, the spectral coefficients $f_{i,j}$ (which may also be viewed as the generalized Fourier coefficients), are given by,

$$f_{i,j} = (f(t), \theta_{i,j}(t)), \forall j \in \mathcal{I}_m, \forall i \in \mathcal{I}_{r_j} \; . \tag{1.9}$$

In practice, the infinite series in (1.8) is truncated, giving us a least squares estimate $\hat{f}(t)$ of the function $f(t)$, i.e.,

$$f(t) \approx \hat{f}(t) = \sum_{j=1}^{m} \sum_{i=1}^{r_j} f_{i,j} \; \theta_{i,j}(t), \tag{1.10}$$

which can be compactly written as

$$\hat{f}(t) = \mathbf{f}^T \boldsymbol{\Theta} \tag{1.11}$$

where,

$$\mathbf{f}^T = [f_{1,1} \cdots f_{r_1,1} | f_{1,2} \cdots f_{r_2,2} | f_{1,m} \cdots f_{r_m,m}] \tag{1.12}$$

and

$$\boldsymbol{\Theta}^T = [\theta_{1,1}(t) \ldots \theta_{r_1,1}(t) | \theta_{1,2}(t) \ldots \theta_{r_2,2}(t) | \theta_{1,m}(t) \ldots \theta_{r_m,m}(t)] \tag{1.13}$$

1.3.3 Completeness

While orthogonality of the set $\mathcal{G}$ leads to many simplifications in the various computational schemes derived using this basis, in order to ensure convergence of $\hat{f}(t)$ to $f(t)$, in the mean with weighting function $\rho(t)$, the completeness of $\mathcal{G}$ must be proved. It is well known that a necessary and sufficient condition for an orthonormal system to be complete is that Parseval's condition holds. That is, in this case, we have to show that,

$$\int_\Omega \rho(t) \, f^2(t) \, dt = \sum_{j=1}^{m} \sum_{i=1}^{\infty} f_{i,j}^2. \tag{1.14}$$

By definition, the set $\mathcal{G}_j$ in Ω_j is complete. Therefore,

$$\int_{\Omega_j} \rho(t) \, f^2(t) \, dt = \sum_{i=1}^{\infty} f_{i,j}^2. \tag{1.15}$$

Now,

$$\int_\Omega \rho(t) \, f^2(t) \, dt = \sum_{j=1}^{m} \int_{\Omega_j} \rho(t) \, f^2(t) \, dt = \sum_{j=1}^{m} \sum_{i=1}^{\infty} f_{i,j}^2 \tag{1.16}$$

in view of (1.15). This proves the completeness of $\mathcal{G}$.

The above results can be summarized in the following statement.

The set $\mathcal{G}$ forms a complete orthogonal set of basis functions in the $\mathbf{L}_{2\rho}$ *space.*

Example 1.1. Approximation of a discontinuous function

The saw-tooth waveform shown in Fig. 1.2 has to be represented in terms of PCBF, CBF and the proposed system of GHOF using only 8 terms over the interval $\Omega = (0,1)$. The three cases considered use the (a) BPF, (b) Legendre Polynomials and (c) GHOF (characterized by $m = 4, r_j = 2, T_j = 0.25, \mathcal{G}_j$ consisting of Legendre polynomials of order 0 and 1, $\forall j \in \mathcal{I}_4$). The vectors of respective spectral coefficients are evaluated using (1.9) (without assuming orthonormality of the basis functions) as

(a) $\mathbf{f}_a^T = [\ 0.25 \quad 0.75 \quad 0.0 \quad 0.0 \quad 0.25 \quad 0.75 \quad 0.0 \quad 0.0 \ ]$

(b) $\mathbf{f}_b^T = [\ 0.25 \ -0.125 \ -0.078 \ -0.164 \ -0.346 \ 0.612 \ 0.266 \ -0.397 \ ]$

(c) $\mathbf{f}_a^T = [\ 0.5 \quad 0.5 \quad 0.0 \quad 0.0 \quad 0.5 \quad 0.5 \quad 0.0 \quad 0.0 \ ]$

The integral of squared error in the approximation is defined to be

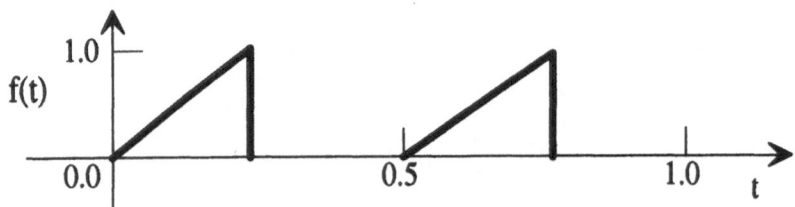

Figure 1.2: The discontinuous function in Example 1.1

$$\mathcal{E}_f = \int_\Omega (f(t) - \hat{f}(t))^2 \; dt. \tag{1.17}$$

The values of $\mathcal{E}_f$ corresponding to the three cases are found to be (a) 0.01, (b) 0.03 and (c) 0.00. Obviously, the GHOF fit (case (c)) is the most natural approximation for the saw-tooth function given here, because of its piecewise continuity mixed with jumps. This basis is chosen with the the segment boundaries coinciding with the points of discontinuities in the function, making it a perfect fit.

Figure 1.3 shows the approximated function $\hat{f}(t)$ corresponding to the three cases (a)–(c), respectively. It is apparent that PCBF or CBF alone will not be suitable for approximation of discontinuous functions of the type given here, which are quite common in electronic circuits.

The new set of GHOF presented in this chapter is seen to be more suitable for approximating functions which are piecewise continuous in nature. This was demonstrated by taking a fixed number (eight) of components in each system of basis functions. In a practical situation, the number of terms in each set may be determined by limiting the error $\mathcal{E}_f$ to a specific value. In the subsequent chapters GHOF will be applied to some problems of systems and control.

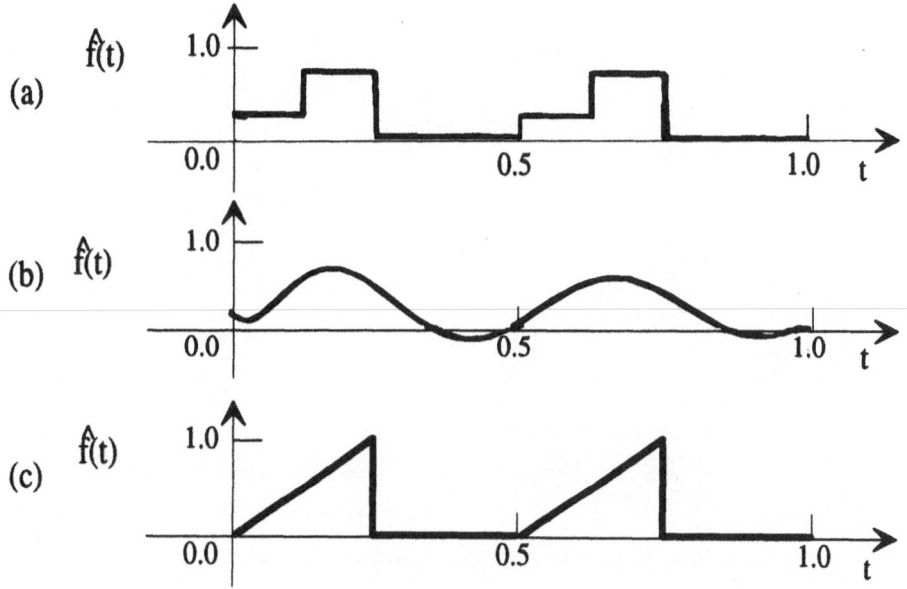

Figure 1.3: Approximations of the discontinuous function in Example 1.1 – using (a) BPF; (b) Legendre polynomials and (c) GHOF

Chapter 2

GHOF Spectral Analysis
of Dynamical Systems

2.1   Survey of literature in the field

In 1973 Corrington applied Walsh functions for the solution of differential and integral equations [64] initiating the activity in the field of orthogonal-functions-based analysis of dynamical systems. This was followed by the work of Chen and Hsiao [32] who used a state space model for analysis. Walsh functions were subsequently applied to the analysis of time-varying systems [43], time-delay systems [45,217], nonlinear systems [164,165,12] etc. Block pulse functions were applied to similar problems at around the same time [248,230,229,231,250,251,257,258]. Application of orthogonal polynomials was initiated in systems and control in the early eighties [112,142] followed by other works [143,194,291]. In the later years there has been an enormous amount of activity in the application of both piecewise constant and continuous basis functions.

To concisely present an overview of the activity in this field, a classified list of literature is given in Table 2.1. Only the system analysis problem is considered here. The types of systems are categorized and abbreviated as defined in "List of Abbreviations" at the beginning of the book. Similarly, the types of basis functions are also denoted by suitable abbreviations. Literature related to distributed parameter systems is not included in this table, since this book deals only with lumped systems.

This chapter is addressed to the problem of analysis of linear and

Table 2.1: Literature map for OF-based analysis

OF used	Types of systems							
	LT	TD	TV	NL	SS	SC	SG	DT
BPF	[36,37,38, 40,109, 117,126, 128,130, 131,167, 191,248, 250,251, 259,274, 290,277]	[39, 144, 188, 229]	[114, 98, 257]	[49, 144, 166, 230, 250, 257, 281]		[34, 114, 231, 218]	[275]	[47, 46, 48]
WF	[32,41,64, 66,177, 189,221, 258,268, 284]	[35, 45, 217, 220]	[43]	[12, 13, 140, 164, 165, 190]		[218]	[14, 169, 195]	[95, 168]
GBPF	[17,288, 289]	[295, 296]		[116]	[17, 99, 297]			
PCBF		[187, 232]	[96]					[97, 96]
OP	[197,271, 272,273]	[172]					[135, 153]	
LAP	[142,256, 294]	[143]				[101, 112]		[118]
LEP	[21,24,196, 290]	[23, 104, 146, 255]	[56, 22, 103, 276]	[56, 106, 252, 255]	[20]	[22, 107, 253]		[119]
CHP	[53,136, 158]	[137, 88, 173]	[156, 149]	[149, 254]		[52, 100]		
JAP	[159]	[90]		[92]		[51]		
HEP	[198]							
GOP	[27,30]	[147, 150]	[148, 151, 278, 292]		[28]	[29, 279]		[151, 279]
SCF	[197,200]	[181]						
GHOF	[204,155]		[154]	[205]				

nonlinear dynamical systems via the system of GHOF. The operational matrix for integration is first derived and used to obtain the solution of a lumped linear system modelled in state space. This is followed by the presentation of a unified framework of numerical analysis where the proposed method is conceptually and numerically compared with the finite element and other orthogonal-functions-based methods. Next, two examples of SCR-controlled DC motor drive systems are analyzed to demonstrate the suitability of GHOF in handling such problems. Finally, the amplitude and time-period of a highly nonlinear van der Pol's oscillator are predicted by a method based on GHOF spectral expansion.

2.2 GHOF operational matrix for integration

It is by now well-known [216] that when the signals are represented in their spectral form, the related operators in continuous-time (CT) domain are reduced to algebraic forms which are approximate in the sense of least squares. For example, the integrator – an operator of extensive use in dynamical systems analysis, can be approximated by the so-called operational matrix for integration. Similarly, other time-domain operations such as differentiation, time-delay, time-scaling etc. can have their corresponding operational matrices. The use of such matrices facilitates systematic development of theory so that techniques based on various systems of orthogonal functions can be brought under a common methodology. This has been separately done in the cases of PCBF [216] and orthogonal polynomials(in the form of generalized orthogonal polynomials, GOP) [25]. However, no attempt has so far been made to unify the somewhat diverse approaches of PCBF and CBF. An attempt, believed to be for the first time, is made here in this direction.

Let $\mathbf{E}_j$ be the operational matrix for integration with respect to $\mathcal{G}_j$. Then by definition,

$$\int_0^\delta \mathbf{p}_j(\tau) \, d\tau \approx \mathbf{E}_j \, \mathbf{p}_j(\delta), \tag{2.1}$$

where,

$$\mathbf{p}_j^T(\delta) = [\, p_{1,j}(\delta/T_j) \ldots p_{r_j,j}(\delta/T_j)\,], \quad \delta \in (0, T_j). \tag{2.2}$$

The approximation in (2.1) is in the sense of least squares. $\mathbf{E}_j$ is a constant, square and invertible matrix which depends on the actual choice of $\mathcal{G}_j$. We define the operational matrix $\mathbf{E}_g$ for $\mathcal{G}$ such that,

$$\int_0^t \Theta(\tau) \, d\tau \approx \mathbf{E}_g \, \Theta(t). \qquad (2.3)$$

Now,

$$\int_0^t \theta_{i,j}(\tau) \, d\tau = \begin{cases} \int_{t_{j-1}}^{t_j} p_{i,j}((\tau - t_{j-1})/T_j) \, d\tau, & t > t_j \\ \int_{t_{j-1}}^{t} p_{i,j}((\tau - t_{j-1})/T_j) \, d\tau, & t \in \Omega_j \\ 0, & t < t_{j-1}. \end{cases} \qquad (2.4)$$

From (2.4), for $t \in \Omega_j$, with $\delta = t - t_{j-1}$ and $\varphi = \tau - t_{j-1}$,

$$\int_0^t \theta_{i,j}(\tau) \, d\tau = \int_0^\delta p_{i,j}(\varphi/T_j) \, d\varphi. \qquad (2.5)$$

Comparing (2.5) with (2.1) and (2.2), and noting that $\mathbf{p}_j$ is the j-th subvector in Θ corresponding to the segment Ω_j, we find that for $t \in \Omega_j$, the l.h.s. of (2.5) may be evaluated using $\mathbf{E}_j$. Therefore, $\mathbf{E}_g$ will be a matrix with diagonal blocks equal to $\mathbf{E}_j, \forall j \in \mathcal{I}_m$. The subdiagonal blocks comprise of $\mathbf{0}$ matrices and the superdiagonal blocks are non-zero and depend on how the first integral in the r.h.s. of (2.4) is expressed in terms of the systems of CBF defined in the succeeding segments $\Omega_{j+1}, \dots \Omega_m$. Therefore, $\mathbf{E}_g$ is a $(\varrho \times \varrho)$ matrix of the following structure.

$$\mathbf{E}_g = \begin{bmatrix} \mathbf{E}_1 & \mathbf{H}_{1,2} & \cdots & \mathbf{H}_{1,m} \\ \mathbf{0} & \mathbf{E}_2 & \cdots & \mathbf{H}_{2,m} \\ \vdots & \vdots & \vdots & \vdots \\ \mathbf{0} & \mathbf{0} & \cdots & \mathbf{E}_m \end{bmatrix}. \qquad (2.6)$$

$\mathbf{E}_j$ is defined by (2.1) and is of order $(r_j \times r_j)$. $\mathbf{H}_{l,j}$ is a $(r_l \times r_j)$ matrix of the form

$$\mathbf{H}_{l,j} = \begin{bmatrix} h_{l,j,1} & & \\ h_{l,j,2} & & \mathbf{0} \\ \vdots & & \\ h_{l,j,r_l} & & \end{bmatrix}, \forall l \in \mathcal{I}_{m-1}, \forall j \in \mathcal{I}_m^{l+1}, \qquad (2.7)$$

where,

$$h_{l,j,i} = \int_{t_{l-1}}^{t_l} \theta_{i,l}(t) \; dt, \forall \, i \in \mathcal{I}_{r_l}. \tag{2.8}$$

It may be noted from (2.8) that $h_{l,j,i}$ is, in fact, independent of j. This fact will be utilized in the next chapter where recursive formulae for calculation of integrals of signals represented using the GHOF basis will be derived.

Among all the systems of orthogonal polynomials, Legendre polynomials have been very widely used in the literature. This system has the weighting function $\rho(t)$ equal to unity making the computation of the spectra quite simple. For this reason, in this book, all numerical examples are worked out with Legendre polynomials. Some results corresponding to this special case are therefore considered now.

The elements of the matrix $\mathbf{H}_{l,j}$ in this case are given by,

$$h_{l,j,i} = \begin{cases} T_l, & i = 1 \\ 0, & i \in \mathcal{I}_{r_l}^2. \end{cases} \tag{2.9}$$

When, in addition, the segments Ω_j are of equal width, i.e.,

$$T_j = T/m, r_j = r, \mathbf{E}_j = \mathbf{E}, \forall j \in \mathcal{I}_m, \mathbf{H}_{l,j} = \mathbf{H}, \forall l \in \mathcal{I}_{m-1}, \forall j \in \mathcal{I}_m^{l+1},$$

then,

$$\mathbf{E}_g = \mathbf{I}_{(m \times m)} \otimes \mathbf{E}_{(r \times r)} + \sum_{i=2}^{m} \mathbf{\Delta}_{(m \times m)}^{i-1} \otimes \mathbf{H}_{(r \times r)}, \tag{2.10}$$

where,

$$\mathbf{\Delta}_{(m \times m)} = \begin{bmatrix} & \mathbf{I}_{(m-1 \times m-1)} \\ \mathbf{0} & \\ & \mathbf{0} \end{bmatrix}, \tag{2.11}$$

and

$$\mathbf{H}_{(r \times r)} = \begin{bmatrix} T/m & \mathbf{0} \\ \mathbf{0} & \end{bmatrix}, \tag{2.12}$$

and the Kronecker product $\otimes$ between two matrices $\mathbf{A}$ and $\mathbf{B}$ is defined as

$$\mathbf{A} \otimes \mathbf{B} = [\, a_{ij}\, \mathbf{B}\,]. \tag{2.13}$$

2.3   Solution of state equation

Let us consider a linear time-invariant system described by

$$\dot{\mathbf{x}}(t) = \mathbf{A}\mathbf{x}(t) + \mathbf{B}\mathbf{u}(t), \ \mathbf{y}(t) = \mathbf{C}\mathbf{x}(t), \tag{2.14}$$

where $\mathbf{x}(t)$, $\mathbf{u}(t)$ and $\mathbf{y}(t)$ are n-vector of state, n_i-vector of inputs and n_o-vector of outputs respectively. $\mathbf{A}$, $\mathbf{B}$ and $\mathbf{C}$ are matrices of appropriate dimensions. Expanding $\mathbf{x}(t)$, $\dot{\mathbf{x}}(t)$ and $\mathbf{B}\mathbf{u}(t)$ in terms of a system of GHOF we obtain,

$$\left.\begin{aligned}
\dot{\mathbf{x}}(t) &\approx \mathbf{V}\boldsymbol{\Theta}(t) \\
\mathbf{x}(t) &\approx \mathbf{X}\boldsymbol{\Theta}(t) \\
\mathbf{B}\mathbf{u}(t) &\approx \mathbf{S}\boldsymbol{\Theta}(t) \\
\mathbf{V} &= [\, \mathbf{v}_{11}\ \mathbf{v}_{21}\ \cdots\ \mathbf{v}_{r_11}\, |\ \cdots\ |\ \mathbf{v}_{1m}\ \mathbf{v}_{2m}\ \cdots\ \mathbf{v}_{r_mm}\,] \\
\mathbf{v}_{ij}^T &= [\, v_{ij1}\ v_{ij2}\ \cdots\ v_{ijn}\,] \\
\mathbf{X} &= [\, \mathbf{x}_{11}\ \mathbf{x}_{21}\ \cdots\ \mathbf{x}_{r_11}\, |\ \cdots\ |\ \mathbf{x}_{1m}\ \mathbf{x}_{2m}\ \cdots\ \mathbf{x}_{r_mm}\,] \\
\mathbf{x}_{ij}^T &= [\, x_{ij1}\ x_{ij2}\ \cdots\ x_{ijn}\,] \\
\mathbf{S} &= [\, \mathbf{s}_{11}\ \mathbf{s}_{21}\ \cdots\ \mathbf{s}_{r_11}\, |\ \cdots\ |\ \mathbf{s}_{1m}\ \mathbf{s}_{2m}\ \cdots\ \mathbf{s}_{r_mm}\,] \\
\mathbf{s}_{ij}^T &= [\, s_{ij1}\ s_{ij2}\ \cdots\ s_{ijn}\,]
\end{aligned}\right\} . \tag{2.15}$$

In the above, v_{ijk}, x_{ijk} and s_{ijk} are the (i,j)-th coefficients of the k-th elements of the vectors $\dot{\mathbf{x}}(t)$, $\mathbf{x}(t)$ and $\mathbf{B}\mathbf{u}(t)$ respectively.

Inserting the above expansion in the state equation (2.14),

$$\mathbf{V}\boldsymbol{\Theta}(t) = \mathbf{A}\mathbf{X}\boldsymbol{\Theta}(t) + \mathbf{S}\boldsymbol{\Theta}(t). \tag{2.16}$$

Since

$$\mathbf{x}(t) = \int_0^t \dot{\mathbf{x}}(\tau)\, d\tau + \mathbf{x}(0), \tag{2.17}$$

we can write

$$\mathbf{X\Theta}(t) = \int_0^t \mathbf{V\Theta}(\tau)\, d\tau + \mathbf{x}_0\mathbf{\Theta}(t) \approx \mathbf{VE}_g\mathbf{\Theta}(t) + \mathbf{x}_0\mathbf{\Theta}(t), \qquad (2.18)$$

where $\mathbf{x}_0$ is the GHOF spectrum of the constant $\mathbf{x}(0)$ and is of the form

$$\mathbf{x}_0 = [\; \underbrace{\mathbf{x}(0)\; \mathbf{0}\; \ldots\; \mathbf{0}}_{r_1}\; |\; \underbrace{\mathbf{x}(0)\; \mathbf{0}\; \ldots\; \mathbf{0}}_{r_2}\; |\; \cdots\; |\; \underbrace{\mathbf{x}(0)\; \mathbf{0}\; \ldots\; \mathbf{0}}_{r_m}\;],$$

since the first term in all the systems of CBF is unity.

Substituting (2.18) in (2.17),

$$\mathbf{V\Theta}(t) = \mathbf{A}\,[\mathbf{VE}_g\mathbf{\Theta}(t) + \mathbf{x}_0\mathbf{\Theta}(t)] + \mathbf{S\Theta}(t),$$

or,

$$\mathbf{V} = \mathbf{AVE}_g + \mathbf{\hat{S}}, \qquad\qquad\qquad\qquad\qquad (2.19)$$

where,

$$\mathbf{\hat{S}} = \mathbf{S} + \mathbf{Ax}_0. \qquad\qquad\qquad\qquad\qquad\qquad (2.20)$$

The effect of the second term in (2.20) is that of a step function. It may be noted that (2.19) is an approximate algebraic relation corresponding to (2.14) which is obtained by making use of the operational matrix for integration $\mathbf{E}_g$.

The explicit solution of (2.19) is given by,

$$\mathcal{V}_{(\varrho n \times 1)} = \left[\mathbf{I}_{(\varrho n \times \varrho n)} - \mathbf{E}_{g(\varrho \times \varrho)} \otimes \mathbf{A}_{(n \times n)}\right]^{-1} \mathcal{S}_{(\varrho n \times 1)}, \qquad (2.21)$$

where,

$$\mathcal{V} = \begin{bmatrix} \mathbf{v}_{11} \\ \mathbf{v}_{21} \\ \vdots \\ \mathbf{v}_{r_1 1} \\ \cdots \\ \mathbf{v}_{1m} \\ \mathbf{v}_{2m} \\ \vdots \\ \mathbf{v}_{r_m m} \end{bmatrix}, \text{ and } \mathcal{S} = \begin{bmatrix} \mathbf{\hat{s}}_{11} \\ \mathbf{\hat{s}}_{21} \\ \vdots \\ \mathbf{\hat{s}}_{r_1 1} \\ \cdots \\ \mathbf{\hat{s}}_{1m} \\ \mathbf{\hat{s}}_{2m} \\ \vdots \\ \mathbf{\hat{s}}_{r_m m} \end{bmatrix}. \qquad (2.22)$$

The state vector is obtained as

$$\mathbf{x}(t) \approx \mathbf{X}\boldsymbol{\Theta}(t) = \mathbf{V}\mathbf{E}_g\,\boldsymbol{\Theta}(t) + \mathbf{x}_0\boldsymbol{\Theta}(t). \tag{2.23}$$

The output $\mathbf{y}(t)$ can now be easily computed.

2.4 Extension of solution beyond the initial interval

Equations (2.21) – (2.23) together constitute the solution of the state equation for the time interval $(0, T)$. However, often it is necessary to obtain the solution beyond this finite interval. It is of course always possible to define the whole domain of interest to be $(0, T)$, but this may require a very large value of ϱ to attain the desired level of accuracy. In such a case, one has to invert a very large matrix in (2.21) with consequent computational complexity and possible numerical difficulties. It is much more advantageous if the value of ϱ can be kept small and the solutions in two successive intervals of time can be linked with appropriate definition of GHOF systems in each interval. This was possible in the case of block pulse functions [221] which led to simple recursive formulations. A similar technique is applied here to recursify the GHOF solution presented above.

The state $\mathbf{x}(t)$ is assumed to be continuous at the boundary of two intervals. The state at the end of the interval $(0, T)$ is given by

$$\mathbf{x}(T) = \mathbf{x}(0) + \sum_{j=1}^{m} \sum_{i=1}^{r_j} \int_{tj-1}^{t_j} \mathbf{v}_{ij}\ \theta_{i,j}(t)\ dt. \tag{2.24}$$

2.4.1 Multiple Segment Multiple Term (MSMT) Formula

To obtain the solution in the adjacent interval $(T, 2T)$ we choose another system of GHOF appropriate to this interval, changing the origin to $t = T$. The state equation can now be solved starting with $\mathbf{x}(T)$ and the new GHOF spectra of the input signal. This procedure can be continued as long as desired. In general, for the k-th interval $((k-1)T, kT)$ we have the following relationship for continuation:

$$
\begin{aligned}
\hat{\mathbf{S}}^{(k)} &= \mathbf{S}^{(k)} + \nu^T \otimes [\mathbf{A}\mathbf{x}((k-1)T)] \\
\mathcal{V}^{(k)} &= \left[\mathbf{I} - \mathbf{E}_g^{(k)^T} \otimes \mathbf{A}\right]^{-1} \mathcal{S}^{(k)} \\
\mathbf{x}(kT) &= \mathbf{x}((k-1)T) + \sum_{j=1}^{m} \sum_{i=1}^{r_j} \int_{t_{j-1}}^{t_j} \mathbf{v}_{ij}^{(k)} \, \theta_{i,j}^{(k)}(t)\, dt \\
\nu^T &= [\underbrace{1\ 0\ \dots\ 0}_{r_1} | \underbrace{1\ 0\ \dots\ 0}_{r_2} | \underbrace{1\ 0\ \dots\ 0}_{r_m}]
\end{aligned}
\quad \bigg\} \quad (2.25)
$$

and the superscript $^{(k)}$ indicates that the quantities belong to the k-th interval. The solution over each interval of time is an approximation in terms of a set $\mathcal{G}^{(k)}$ on multiple segments and over each segment $\Omega_j^{(k)}$, $\mathcal{G}_j^{(k)}$ contains multiple terms. Thus we may refer to (2.25) as multiple segment multiple term (MSMT) formula. This is the most general type of approximation under the GHOF framework. Several special cases will now be presented mentioning the advantages and limitations of each type.

2.4.2   Single Segment Multiple Term (SSMT) Formula

As already mentioned, the solution via (2.21) involves the inversion of a matrix of size ($\varrho n \times \varrho n$). Therefore, sometimes it may become impracticable to use this formula, especially when m, and consequently ϱ, is large. In order to avoid this difficulty, the solution may first be obtained for $m = 1$ with a proper choice of T_1. Thereafter the solution may be extended to any length of time by using the technique of extension given above. There need not be any restriction on either the widths of the segments T_k, the type of CBF chosen in each segment or the number of orthogonal components r_k. In this way, it is possible to retain the full generality of the definition of GHOF while obtaining a simplified computational scheme. Equation (2.25) in such situations reduces to:

$$
\begin{aligned}
\hat{\mathbf{S}}_*^{(k)} &= \mathbf{S}_*^{(k)} + \nu_k^T \otimes [\mathbf{A}\mathbf{x}(t_{k-1})] \\
\mathcal{V}_*^{(k)} &= \left[\mathbf{I} - \mathbf{E}_k^T \otimes \mathbf{A}\right]^{-1} \mathcal{S}_*^{(k)} \\
\mathbf{x}(t_k) &= \mathbf{x}(t_{k-1}) + \sum_{i=1}^{r_k} \int_0^{T_k} \mathbf{v}_{i1}^{(k)} \, \theta_{i,1}^{(k)}(t)\, dt \\
t_k &= \sum_{l=1}^{k} T_l, \nu_k^T = [\underbrace{1\ 0\ \dots\ 0}_{r_k}]
\end{aligned}
\quad \bigg\} \quad (2.26)
$$

The subscript ∗ in the above indicates that the quantities correspond to the situation when $m = 1$. Equation (2.26) can be viewed as a recursive continuation formula for CBF solutions and will be referred to as the single segment multiple term (SSMT) formula.

2.4.3 Multiple Segment Single Term (MSST) Formula

If we retain only the first term in $\mathcal{G}_j, \forall j \in \mathcal{I}_m$, i.e., $r_j = 1, \forall j \in \mathcal{I}_m$ and further set $T_j = T/m, \forall j \in \mathcal{I}_m$, we obtain the well known PCBF (the BPF expansion in particular) formula [216].

$$
\begin{aligned}
\hat{\mathbf{S}}_{\ast\ast}^{(k)} &= \mathbf{S}_{\ast\ast}^{(k)} + \kappa^T \otimes [\mathbf{Ax}((k-1)T)] \\
\mathcal{V}_{\ast\ast}^{(k)} &= \left[\mathbf{I} - \mathbf{E}_b^T \otimes \mathbf{A}\right]^{-1} \mathcal{S}_{\ast\ast}^{(k)} \\
\mathbf{x}(kT) &= \mathbf{x}((k-1)T) + T/m \sum_{j=1}^{m} \mathbf{v}_{1j}^{(k)} \\
\kappa^T &= \underbrace{\begin{bmatrix} 1 & 1 & \dots & 1 \end{bmatrix}}_{m}
\end{aligned} \quad \Bigg\} . \tag{2.27}
$$

In the above, the subscript ∗∗ signifies the special case of PCBF (BPF) formula *en bloc* (i.e., with multiple segments) and $\mathbf{E}_b$ is the operational matrix for integration via BPF. The well established PCBF formula is thus seen to be the special case of multiple segment single term (MSST) formula of the GHOF scheme.

2.4.4 Single Segment Single Term (SSST) Formula

Finally, the simplest case of $m = 1$ and $r_1 = 1$ is considered. This leads to the well-known recursive formula via BPF [221].

$$
\begin{aligned}
\mathbf{V}_{\ast\ast\ast}^{(k)} &= [\mathbf{I} - \mathbf{A}T/2]^{-1}\left[\mathbf{S}_{\ast\ast\ast}^{(k)} + \mathbf{Ax}((k-1)T)\right] \\
\mathbf{X}_{\ast\ast\ast}^{(k)} &= T/2\,\mathbf{V}_{\ast\ast\ast}^{(k)} \\
\mathbf{x}(kT) &= \mathbf{x}((k-1)T) + T\mathbf{V}_{\ast\ast\ast}^{(k)}
\end{aligned} \quad \Bigg\} . \tag{2.28}
$$

Here $\mathbf{V}_{\ast\ast\ast}^{(k)}$, $\mathbf{X}_{\ast\ast\ast}^{(k)}$ and $\mathbf{S}_{\ast\ast\ast}^{(k)}$ are n-vectors. The recursive BPF formula, also termed as single segment single term (SSST) formula, is obtained over successive time segments in successive intervals. If this approximation is allowed, the solution becomes computationally most efficient.

Ideally, in terms of accuracy, the MSMT and SSMT approximations are equivalent. Likewise, the MSST and SSST approximations are equivalent. However, from points of view of computational efficiency and susceptibility to numerical errors, the recursive formulae employing single segment expansions are more efficient than those having multiple segments. In summary, it should be remembered that the accuracy increases with the number of terms in the respective segments, while recursification simplifies computation.

2.5 General framework of numerical analysis of dynamical systems

At this stage it would be interesting to attempt to form a general framework of numerical analysis unifying these orthogonal functions (OF)-based techniques with several other techniques which have already been unified under the name of finite element (FE) methods [312]. The FE methods have been widely used, particularly in the context of distributed parameter systems, i.e., in the solution of partial differential equations. However, as far as the authors are aware, the FE and OF techniques have not been visualized so far as being cases of a common methodology. The purpose of this section is to bring out this underlying unity.

The general procedure for obtaining approximate solutions of differential equations encompassing the two classes of methods is first outlined and the particular conditions which lead to a specific technique are pointed out. Then a simple example of a lumped linear system excited by a discontinuous forcing function is illustrated.

Let us consider a system described by the general differential equation

$$\mathcal{D}\left(\mathbf{y}(t), \mathbf{u}(t), t\right) = 0, \tag{2.29}$$

where, $\mathbf{u}(t)$ and $\mathbf{y}(t)$ are the input and output signals of the system. The problem is to determine $\mathbf{y}(t), \forall t \in \Omega \subset \mathbf{R}_1$, given the values of $\mathbf{u}(t), \forall t \in \Omega$, subject to a set of boundary (or initial) conditions

$$\mathcal{B}\left(\mathbf{y}(t), \mathbf{u}(t), t\right) = 0 \text{ on boundary } \Gamma. \tag{2.30}$$

To solve this problem we often replace (2.29) with a suitable integral equation (as in the case of OF methods),

$$\mathcal{I} = \int \cdots \int \mathcal{D}\left(\mathbf{y}(t), \mathbf{u}(t), t\right) \, dt^l = 0, \qquad (2.31)$$

Further, to make the problem numerically tractable, the continuous-time functions are assumed in the form

$$\mathbf{y} \approx \hat{\mathbf{y}} = \mathbf{Y}\,\Theta \qquad (2.32)$$

and

$$\mathbf{u} \approx \hat{\mathbf{u}} = \mathbf{U}\,\Theta, \qquad (2.33)$$

where Θ is a vector of basis functions defined over Ω. This may consist of orthogonal or non-orthogonal polynomials, sine-cosine functions, trains of impulses etc.

With approximations (2.32) and (2.33), the l.h.s. of (2.31) may be replaced by the approximate integral form

$$\mathcal{A} = \int \cdots \int \mathcal{D}\left(\hat{\mathbf{y}}(t), \hat{\mathbf{u}}(t), t\right) \, dt^l. \qquad (2.34)$$

Again, the domain Ω is often divided into elements Ω_j such that $\Omega = \bigcup_j \Omega_j$ with corresponding boundaries Γ_j. Over every element Ω_j, we may obtain the solution by one of the following two approaches:

A. *Weighted residual method*

We set

$$\int_{\Omega_j} \boldsymbol{\Psi}^T \mathcal{A} \, d\Omega + \int_{\Gamma_j} \boldsymbol{\Psi}'^T \mathcal{B} \, d\Gamma = 0, \qquad (2.35)$$

where $\boldsymbol{\Psi}$ and $\boldsymbol{\Psi}'$ are suitable weighting functions which may be polynomials, impulse trains etc.

B. *Minimization of a functional*

We define a functional

$$\mathcal{J} = \int_{\Omega_j} f(\mathcal{A}) \, d\Omega + \int_{\Gamma_j} f'(\mathcal{B}) \, d\Gamma, \qquad (2.36)$$

where $f(\cdot)$ and $f'(\cdot)$ are suitably chosen error functions. We then minimize $\mathcal{J}$ with respect to the unknowns in (2.32).

Within the above framework, some of the commonly used schemes along with the proposed GHOF technique are categorized in Table 2.2. The FE methods do not involve any initial integration of the differential equation $\mathcal{D}$ (characterized by $l = 0$) while, in all the OF-based schemes, $\mathcal{D}$ is integrated n-times where n is the order of $\mathcal{D}$. This is done to facilitate the use of the operational matrix for integration. The vector Θ is termed as the shape function in FE literature and may consist of either orthogonal or non-orthogonal functions. On the other hand, in the OF-based methods, Θ always consists of orthogonal basis functions and the methods derive their names from the specific functions used. While both the approaches A and B are fairly common in the FE literature, OF methods so far have used only approach A. Depending on the choice of the weighting functions Ψ and Ψ', in approach A, several variants of the FE methods are available, such as the Galerkin method (1(a)), finite difference methods (forward, central or backward difference, 1(b)), subdomain collocation method (1(c)) etc. The Galerkin method is characterized by $\Psi = \Theta$ and this often leads to a formulation equivalent to the variational method (1(d), approach B) where $\mathcal{J}$ is defined to be the variational functional of the differential equation under consideration [312]. The least squares method (1(e)) is another variant of the FE method in which the integral of squared error in the approximation (2.32) is minimized. In all the OF schemes the weighting function Ψ is chosen to be the same as the basis function Θ, implying that the methods are of the Galerkin type. In the BPF (2(a)) and the GHOF (2(b)) methods, the domain Ω is divided into smaller elements Ω_j, as in the FE schemes. Further, these two methods use orthogonal functions which are defined separately over each of these elements in a disjointed manner leading to considerable simplifications in the resulting formulations. While the BPF method always approximates the signals in a piecewise constant form, the GHOF allow higher order continuous approximation over each element. In this sense the GHOF method is as general as the FE methods.

Example 2.1. Analysis of a lumped linear system with a discontinuous input signal

Let us consider a first order differential equation

Table 2.2: OF and other approaches in a common framework of numerical analysis

Scheme	Name of the method	l	Approach	Θ	Ψ, Ψ'	$f(\cdot), f'(\cdot)$	Ω
Finite element methods							
1(a)	Galerkin Method	0	A	Poly or SCF	Θ	–	$\bigcup_j \Omega_j$
1(b)	Finite difference methods	0	A	Poly or SCF	Impulses	–	$\bigcup_j \Omega_j$
1(c)	Subdomain collocation method	0	A	Poly or SCF	Constant	–	$\bigcup_j \Omega_j$
1(d)	Variational method	0	B	Poly or SCF	–	Variational	$\bigcup_j \Omega_j$
1(e)	Least squares method	0	B	Poly or SCF	–	$(\cdot)^2$	$\bigcup_j \Omega_j$
Orthogonal functions methods							
2(a)	BPF	n	A	BPF	Θ	–	$\bigcup_j \Omega_j$
2(b)	GHOF	n	A	GHOF	Θ	–	$\bigcup_j \Omega_j$
2(c)	CBF	n	A	CBF	Θ	–	Ω_j

$$\dot{y} + ay - au = 0, \tag{2.37}$$

where $u(t)$ driving the system is a saw-tooth signal like the one used in Example 1.1 (Fig. 1.2). The input signal is approximated with 10-element linear *shape* functions of the type shown in Fig. 2.1. The approximated signal is shown in Fig. 2.2(a). Fig. 2.2(b) shows the corresponding approximations using the BPF method, GHOF method (with 2 segments and 5 Legendre polynomial terms over each segment) and a 10-term polynomial approximation using Legendre polynomials. These three schemes are referred to as OF schemes 1, 2 and 3 respectively. It may be observed that the GHOF approximation of the input signal is the most accurate since the discontinuity in the input signal at $t = 0.5$ is naturally modelled by this basis. The slight mismatch between the actual and the approximated signals in this case is due to numerical errors in the computation of the higher order spectral components. These three OF schemes and five FE schemes are taken up for a comparative study. The first four FE schemes are based on approach A and the weighting functions used are shown in Fig. 2.1. These schemes are commonly known as the central difference method, backward difference method and Galerkin method (Galerkin – 1 & 2 respectively). The fifth FE scheme is the least squares scheme [312]. These schemes are labelled as FE schemes 1 to 5. The forward difference scheme is seen to be unstable in the present case and is therefore not considered. The subdomain collocation method and the variational method coincide with FE schemes 1 and 3 respectively. Further details regarding these schemes are available in [312].

Solutions are obtained for various values of the parameter a, and the results are summarized in Table 2.3. The normalized integral of squared error in the output is defined to be

$$\mathcal{E}_y = \frac{\int_0^{1.0} (y(t) - \hat{y}(t))^2 \, dt}{\int_0^{1.0} y^2 \, dt}, \tag{2.38}$$

and is evaluated numerically. It may be observed from Table 2.3 that while all methods perform well when the system bandwidth is small, the GHOF method (OF scheme 2) retains its efficiency for all values of a. This is again due to the the capability of the GHOF basis to represent both the continuity and discontinuity in the signals. The output waveforms obtained using the various methods are shown in Figs. 2.3 and 2.4. and these graphs clearly bring out the superiority of the GHOF method over the other schemes considered here.

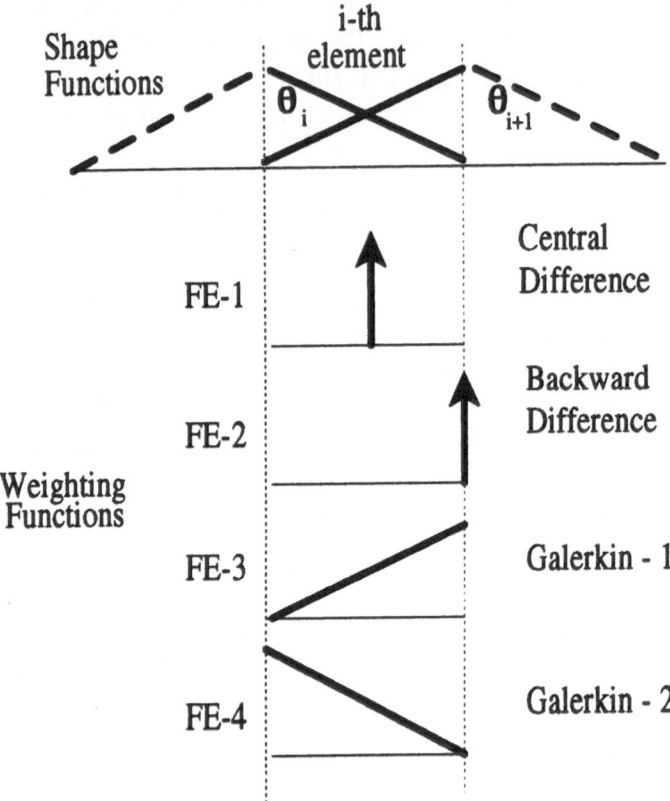

Figure 2.1: Shape and weighting functions for finite element (FE) approximation

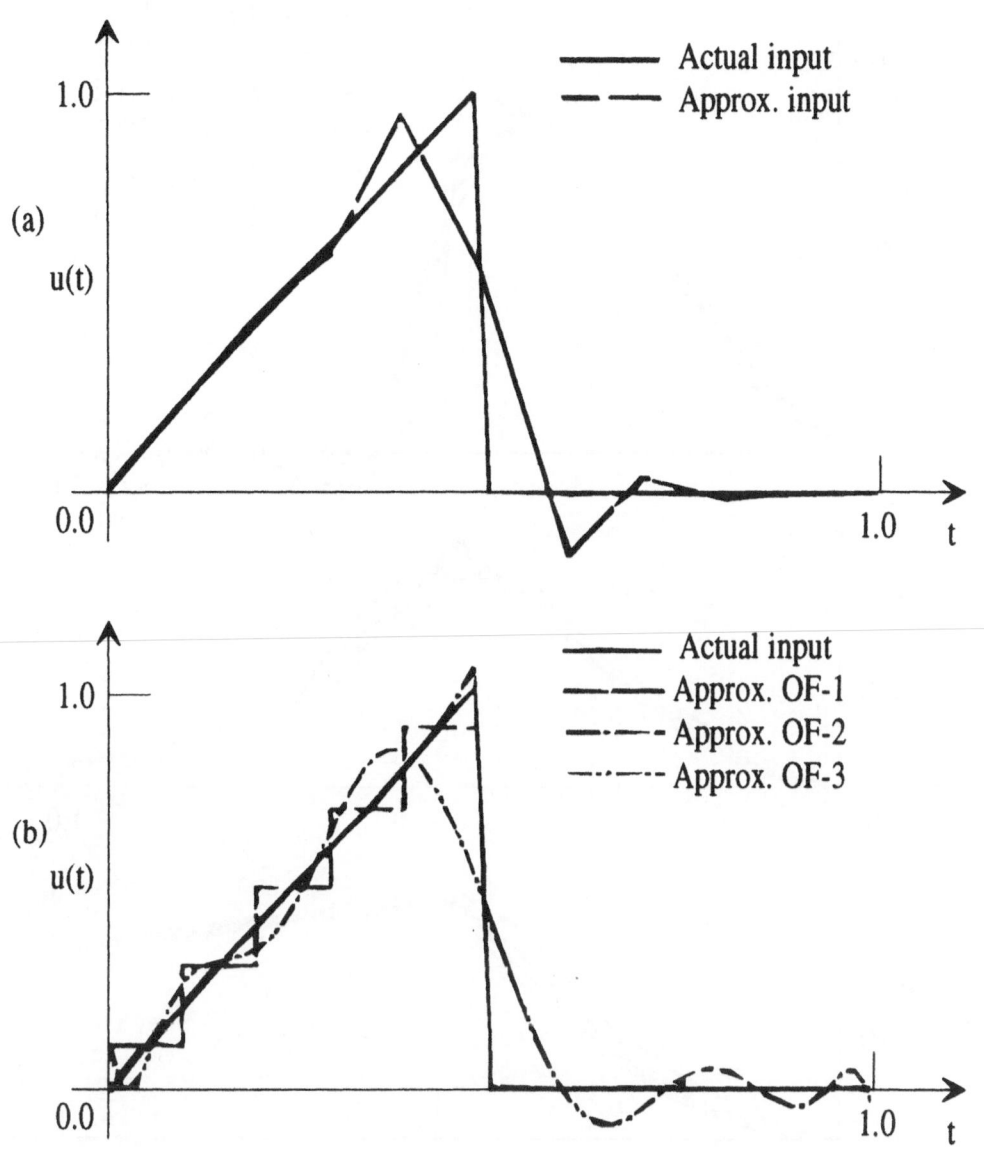

Figure 2.2: Approximation of input signal in Example 2.1 by (a) FE schemes and (b) OF schemes

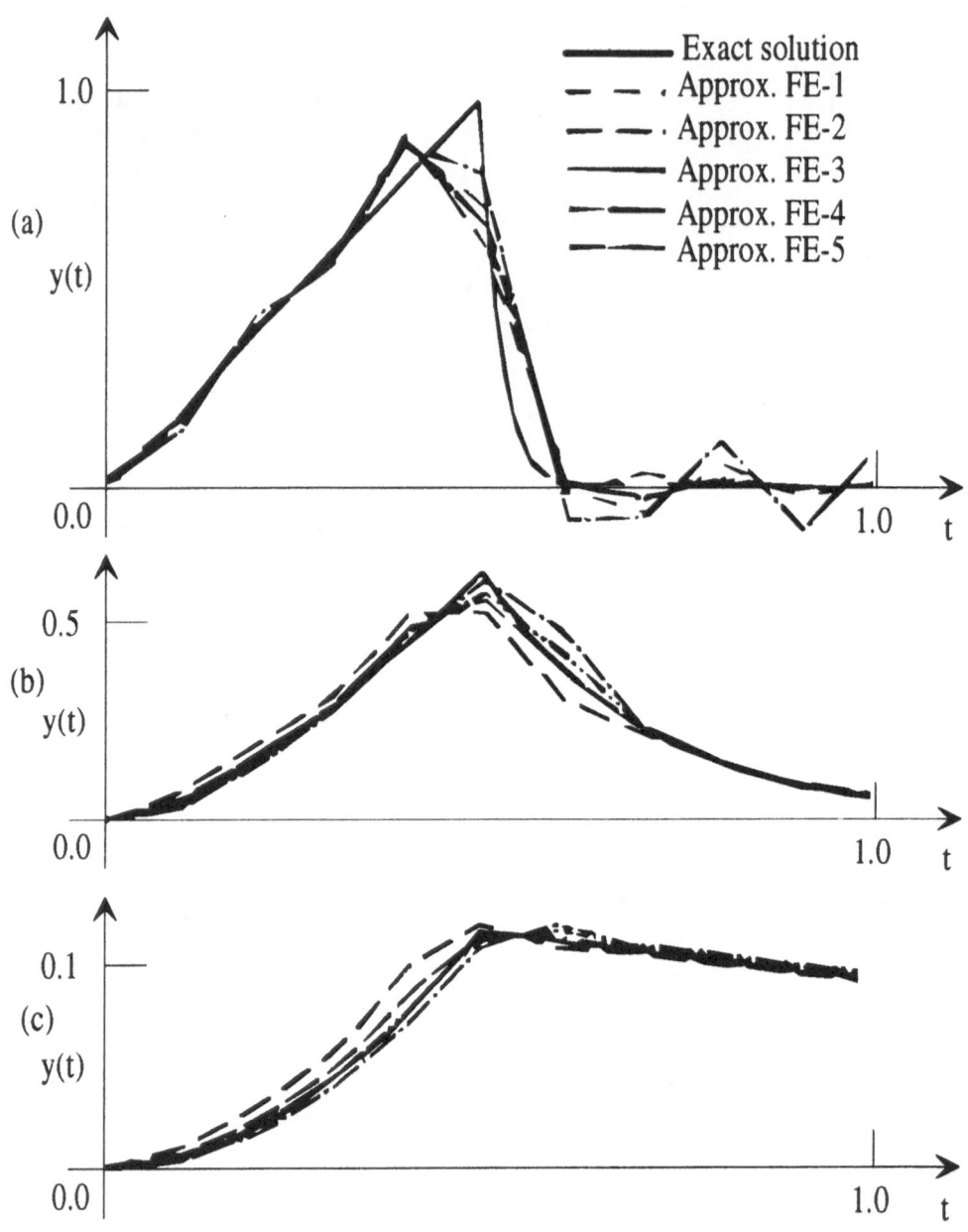

Figure 2.3: Solutions of Example 2.1 using FE schemes for (a) $a = 100$, (b) $a = 10$ and (c) $a = 1$

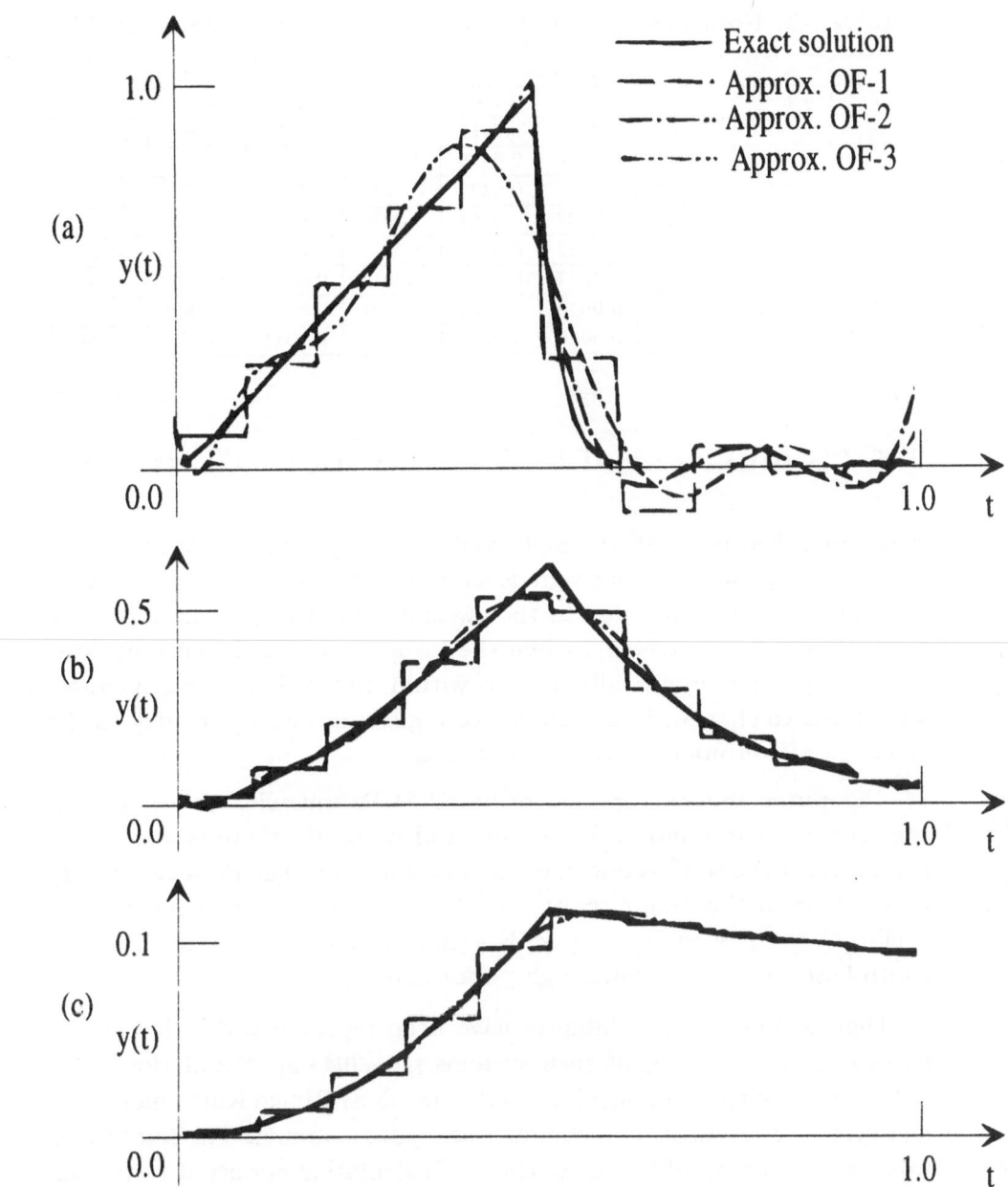

Figure 2.4: Solutions of Example 2.1 using OF schemes for (a) $a = 100$, (b) $a = 10$ and (c) $a = 1$

Table 2.3: Normalized integral of squared errors $\mathcal{E}_y$ in Example 2.1

a	FE methods					OF methods		
	FE–1	FE–2	FE–3	FE–4	FE–5	OF–1	OF–2	OF–3
100.0	0.0421	0.0410	0.0407	0.0484	0.0409	0.0588	0.0109	0.0526
50.00	0.0230	0.0250	0.0214	0.0329	0.0216	0.0532	0.0026	0.0320
20.00	0.0082	0.0185	0.0076	0.0172	0.0073	0.0306	0.0001	0.0110
10.00	0.0037	0.0169	0.0037	0.0107	0.0032	0.0178	0.0000	0.0042
5.000	0.0019	0.0145	0.0022	0.0068	0.0017	0.0101	0.0000	0.0018
2.000	0.0011	0.0115	0.0017	0.0038	0.0011	0.0057	0.0000	0.0008
1.000	0.0009	0.0110	0.0018	0.0027	0.0009	0.0046	0.0000	0.0006
0.500	0.0009	0.0113	0.0020	0.0022	0.0009	0.0041	0.0000	0.0005

2.6 Simulation of SCR-controlled DC drives

A common feature of all the SCR-controlled drive systems is the occurrence of jumps in the input signals as well as in the system parameters. The former takes place due to the switching of the SCRs according to a predefined firing strategy, while the latter occurs if the current flowing through the circuit falls to zero within one cycle of input voltage. This leads to changes in the circuit configurations and gives rise to the so-called *discontinuous mode* of operation.

Computer and microprocessor-based SCR-controlled drives are extensively used to control DC motors and presently there is a trend to extend the same to the control of AC motors [152]. The most commonly used power modulating circuits are converters, choppers, inverters and cycloconverters of various types. Nowadays, they incorporate advanced control strategies to achieve high performance [152,186].

Digital simulation techniques have been reported widely in the literature for the analysis of such systems [185,304,65]. Simulation algorithms employing numerical integration (such as Runge Kutta methods) and system discretization techniques are quite common. The samples of input signals are used for the purpose of calculation of output responses such as speed and current. Normally a large number of samples is required to approximate the irregular type of waveforms and the choice of the integration step length is also very critical. Application of orthogonal functions is not very common since the conventional Fourier and polynomial spectra are unwieldy to handle in view of the large number of terms required. However, recently a method using Walsh functions has been applied to the analysis of a chopper system [66].

The framework of GHOF is appropriate in these situations. The piecewise continuous nature of the input voltages and the resulting current and speed waveforms of the motor can be efficiently approximated using this basis. The jumps in the system parameters can be handled in a straightforward way by choosing different segments for each situation and employing the single segment multiple term (SSMT) formula which can be recursively continued as long as desired. The width of each segment and the number of terms in each segment can be chosen such that the computational effort is minimized without sacrificing accuracy.

Two examples are considered here to demonstrate the efficiency of GHOF as a new tool for the above problems. Both the examples deal with open-loop armature speed control of a DC motor with a separately excited constant field. The power modulation circuit in one case is a 3-phase converter and in the other, a DC chopper.

The DC motor is modelled by standard state space equations ignoring mechanical friction. In this model, the various entities are as follows:

$$
\left.
\begin{aligned}
\mathbf{x}^T \;\; &= \;\; \mathbf{y}^T = [\text{ motor current } i(t),\ \text{motor speed } \omega(t)] \\[2mm]
\mathbf{u}^T \;\; &= \;\; [\text{ line voltage } v_L(t),\ \text{load torque } T_L] \\[2mm]
\mathbf{A} \;\; &= \;\; \begin{bmatrix} -R/L & -K_a/L \\ K_T/J & 0 \end{bmatrix} \\[4mm]
\mathbf{B} \;\; &= \;\; \begin{bmatrix} 1/L & 0 \\ 0 & -1/J \end{bmatrix} \\[4mm]
\mathbf{C} \;\; &= \;\; \begin{bmatrix} 1 & 0 \\ 0 & 1 \end{bmatrix}
\end{aligned}
\right\} , \quad (2.39)
$$

where R and L are the resistance and inductance of the motor armature. K_a and K_T are the armature voltage and torque constants and J is the moment of inertia. The model (2.39) is valid when the SCR-s are *on*, i.e., current is flowing into (or from) the DC motor. During the *off* period, the equation relating the electrical quantities is no longer relevant such the the system model changes to

$$\mathbf{u}^T = [0, \text{ load torque } T_L]$$

$$\mathbf{A} = \begin{bmatrix} 0 & 0 \\ 0 & 0 \end{bmatrix}$$

$$\mathbf{B} = \begin{bmatrix} 1/L & 0 \\ 0 & -1/J \end{bmatrix} \qquad (2.40)$$

$$\mathbf{C} = \begin{bmatrix} 1 & 0 \\ 0 & 1 \end{bmatrix}$$

For simplicity, parameters such as voltage drop across SCRs, commutation intervals, source impedance etc. are neglected here.

Example 2.2. Converter driven DC motor

Consider the 3-phase fully controlled bridge converter system shown in Fig. 2.5. By firing the SCRs in a suitable sequence, a rectified voltage is available at the motor terminals. Once an SCR is fired it will remain in the *on* state until it is forced to the *off*-state by line commutation (switching of the next SCR in sequence) or by reduction of motor current to zero, whichever is earlier. The second situation gives rise to the *discontinuous mode* of operation which occurs if the extinction angle β (corresponding to the time instant when the current falls to zero) satisfies the inequality

$$\beta < \alpha + \pi/3, \qquad (2.41)$$

where α is the firing angle (with α and β measured from a reference point, which is 60° ahead of the natural commutation point). For a given α, the value of β can be calculated by solving the following transcendental equation [185]

$$\cos\phi\sin(\beta-\phi)-k_n+e^{-(\beta-\alpha)/\tan\phi}[k_n-\cos\phi\sin(\alpha-\phi)] = 0, \quad (2.42)$$

where $\phi = \tan^{-1}(\omega_s L/R), \omega_s = $ supply frequency, and $k_n = K_a\omega/V_L$, $V_L = $ being the peak line voltage.

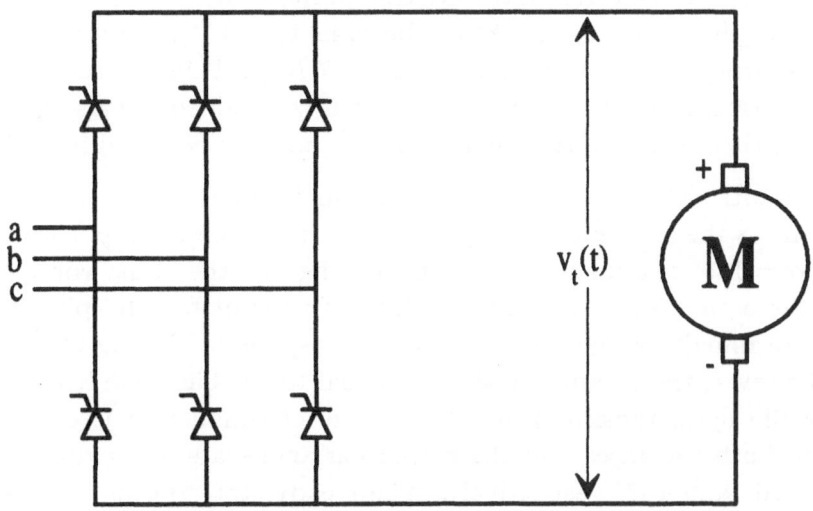

Figure 2.5: The Converter-driven DC motor system of Example 2.2

In the general case of *discontinuous mode*, the input signal per cycle can be approximated by a GHOF of $m = 2$. The time intervals T_1 and T_2 correspond to the *on* and *off* periods respectively. T is equal to one *cycle*, which is $\frac{1}{6}$th of the time period of the AC supply voltage. In the *continuous mode* of operation, $m = 1$ and the second interval disappears.

The SSMT formula given by (2.26) is used to simulate a system with the following motor parameters [185]:

$$V_{\text{rated}} = 400\text{V}, \ i_{\text{rated}} = 137.5\text{A}, \ \omega_{rated} = 157.1\text{rad/s}$$

$$R = 0.097164\Omega, \ L = 0.0055675\text{H}, \ J = 1.7522\text{Kg-m}^2$$

$$K_a = 2.4611\text{V-s/rad}, \ K_T = 2.5470\text{N-m/A}.$$

The simulation uses Legendre polynomials over both the segments. The first three terms of Legendre set are found to describe efficiently the chopped waveform of $v_L(t)$. During the off-period all system variables remain nearly constant. Therefore, it is sufficient to model these variables by the first Legendre term alone. Since the order of the system is 2, for efficient representation of integrated signals, we add this to the number of terms giving $r_1 = 3 + 2 = 5, r_2 = 1 + 2 = 3$. The number of samples used for calculation of the GHOF spectra of the input voltage is 200. The supply voltage v_L is taken to be equal to 380 V(r.m.s.) and the supply frequency is 60 Hz. T_L is equal to 27.7 N-m. Thus the setting is exactly identical to that in [185] and the results obtained are also quite similar. Further details of the simulation program can be found in [205].

The results of the simulation are shown in Figures 2.6 and 2.7. Fig. 2.6 shows the steady state response over three cycles of input waveform for $\alpha = 120°$ and for $\alpha = 110°$. In these figures, the "Line Voltage" $v_L(t)$ is the actual input voltage that is fed to the motor. It coincides with the terminal voltage $v_t(t)$ during the on-period. During the off-period, however, the terminal voltage is equal to the back e.m.f. of the motor, while $v_L(t)$ remains zero. The mode of conduction is discontinuous in both the cases and the extinction angles are approximately equal to 170° and 167° respectively. The steady state speeds at these two operating points are found out to be 120.7 rad/s and 137.9 rad/s while the theoretically predicted values (based on approximate analytical methods) are 120.74 rad/s and 138.6 rad/s respectively. The results agree quite well with the simulations done in [185,65]. Fig. 2.7 shows the transient response of the system when the firing angle is suddenly changed from 120° to 110°. These results are also similar to [185,65] which had been obtained using elaborate digital simulation techniques.

Example 2.3. Chopper driven DC motor

Consider the system shown in Fig. 2.8. A chopper drives a DC motor which has a free-wheeling diode (FWD) connected across it. For simplicity, the commutation circuit is not shown in detail. The discontinuous mode of operation sets in if the motor current falls to zero before the off-interval of the chopper is over, as shown in Fig. 2.9. In the continuous mode of operation, every cycle of input voltage is characterized by $m = 2$ with the two time intervals T_1 and T_2 corresponding to T_{ON} and T_{OFF} respectively. In the discontinuous mode, T_{OFF} is further divided into two subintervals T_2 and T_3 which denote the free-wheeling interval and the zero-current interval respectively.

As in Example 2.2, the system is represented by equation (2.39) when the current is not zero and by (2.40) when it is zero. The currents at the instants $t = T_1$ and $t = T_1 + T_2$ are related by

$$i(T_1 + T_2) = -(K_a\omega/R)(1 - e^{-T_2 R/L}) + i(T_1)e^{-T_2 R/L} \qquad (2.43)$$

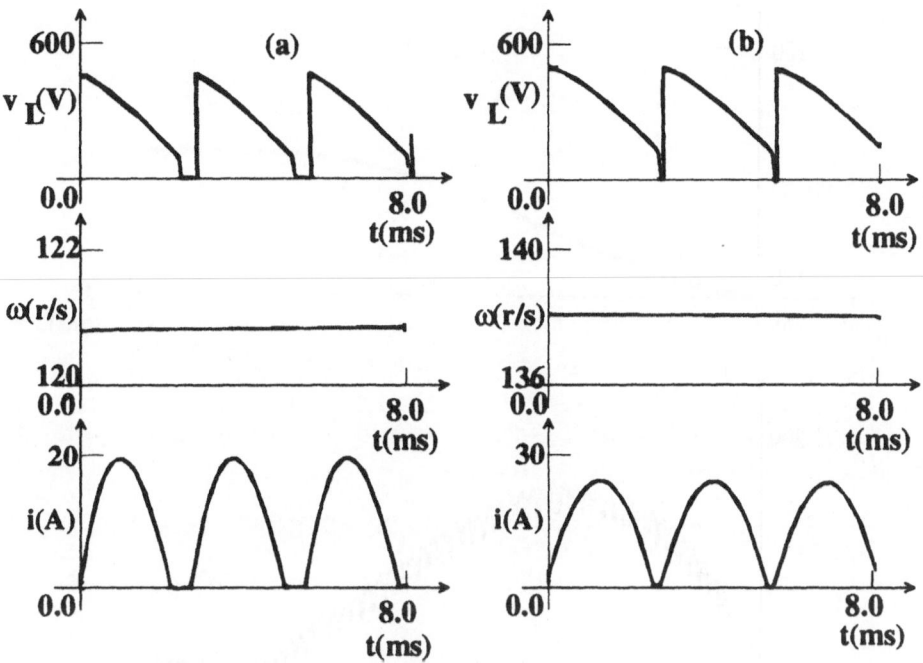

Figure 2.6: Steady state waveforms in Example 2.2 for (a) $\alpha = 120°$ and (b) $\alpha = 110°$

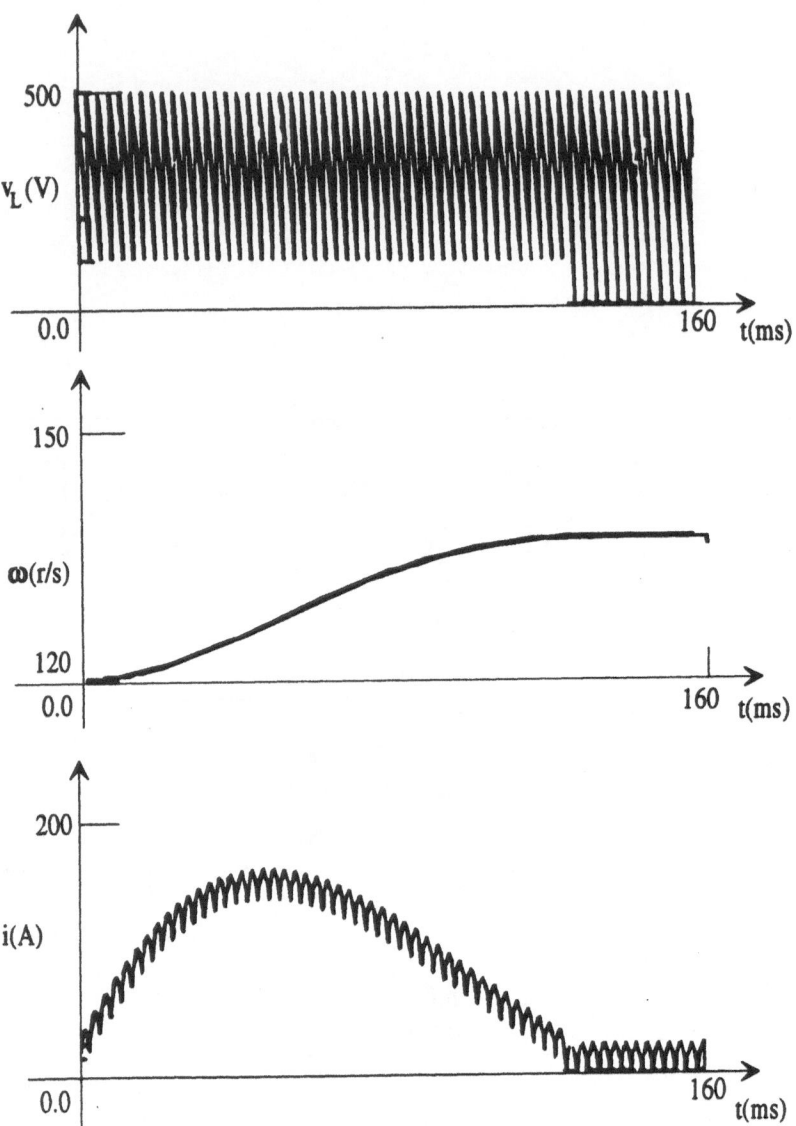

Figure 2.7: Transient response in Example 2.2 for a sudden change in α from 120° to 110°

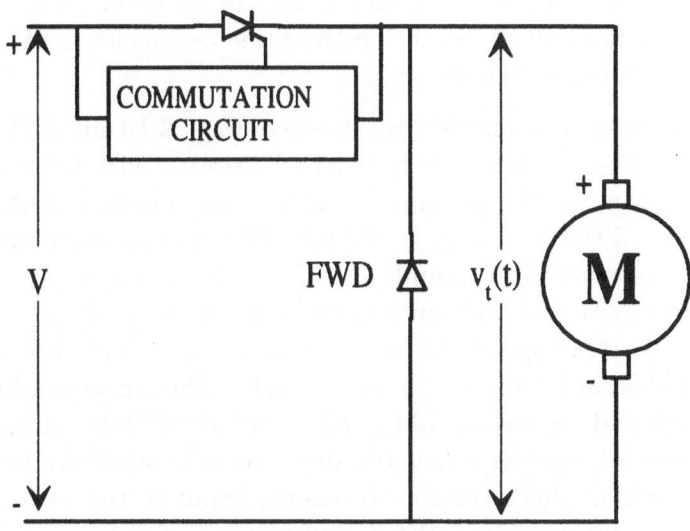

Figure 2.8: The chopper-driven DC motor system of Example 2.3

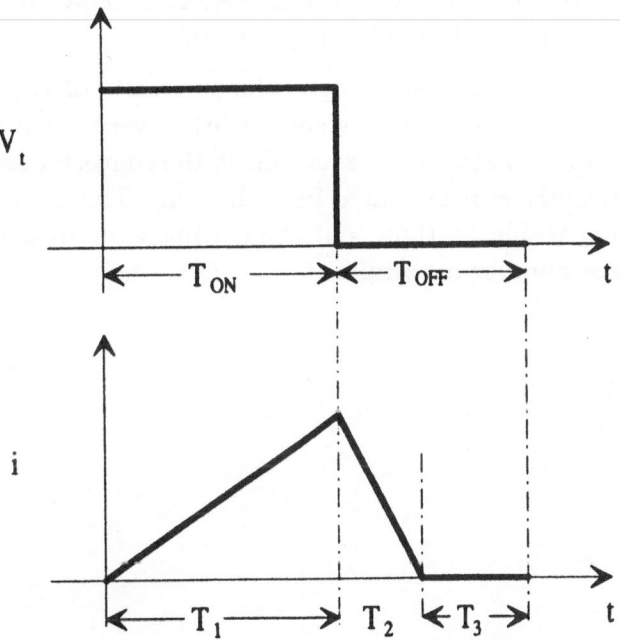

Figure 2.9: Voltage and current waveforms in the discontinuous mode for the system of Example 2.3

Given $i(T_1)$ and assuming ω to be constant over the interval T_2, equation (2.43) may be used to test for the mode of conduction. In the discontinuous mode of operation the same can be solved for T_2, and T_3 can be obtained therefrom.

The results of simulation are shown in Figs. 2.10 and 2.11. The values of r_1, r_2 and r_3 are chosen to be 3, 3 and 2 respectively for reasons similar to those given in the previous example. The supply voltage is taken to be 400 V and the frequency is 200 Hz. All the other input variables have the same values as in Example 2.2. Fig. 2.10 shows the steady state responses for 50% and 75% duty cycles (the ratio $\frac{T_{ON}}{T_{ON}+T_{OFF}}$) respectively. The steady state speeds at the two operating points are found to be 129.98 rad/s and 146.0 rad/s respectively. The corresponding theoretically predicted values are 130.4 rad/s and 146.3 rad/s. Fig. 2.11 shows the transient response when the duty cycle is suddenly changed from 50% to 75%. In this situation the energy input to the motor is very low because the current is mostly discontinuous and it takes an appreciably large amount of time for the motor to reach the new steady state operating point. For this reason only the average values of voltage, speed and current over every cycle are shown in Fig. 2.11. The CBF chosen in this example also are the Legendre polynomials.

Examples 2.2 and 2.3 demonstrate the relevance of the GHOF in the analysis of power electronic systems. With a very small number of terms (about 8) in each case, the waveforms in the converter and chopper circuits are efficiently approximated in each cycle. The accuracy of the solutions is comparable to those obtained using numerical techniques with a very large number of samples.

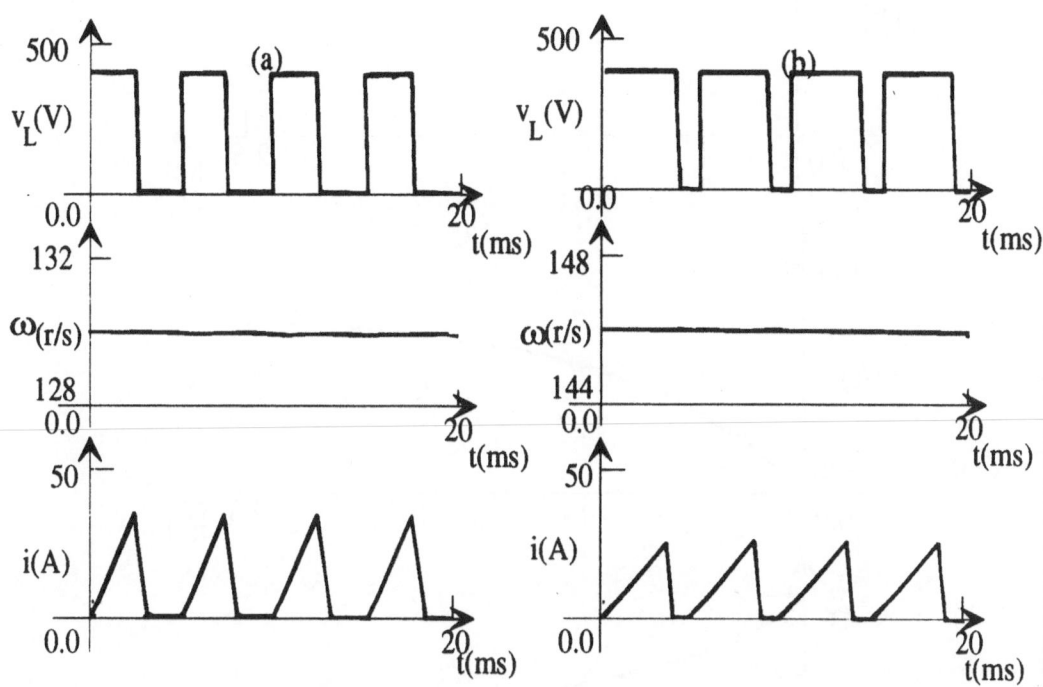

Figure 2.10: Steady state waveforms in Example 2.3 for duty cycle of (a) 50% and (b) 75%

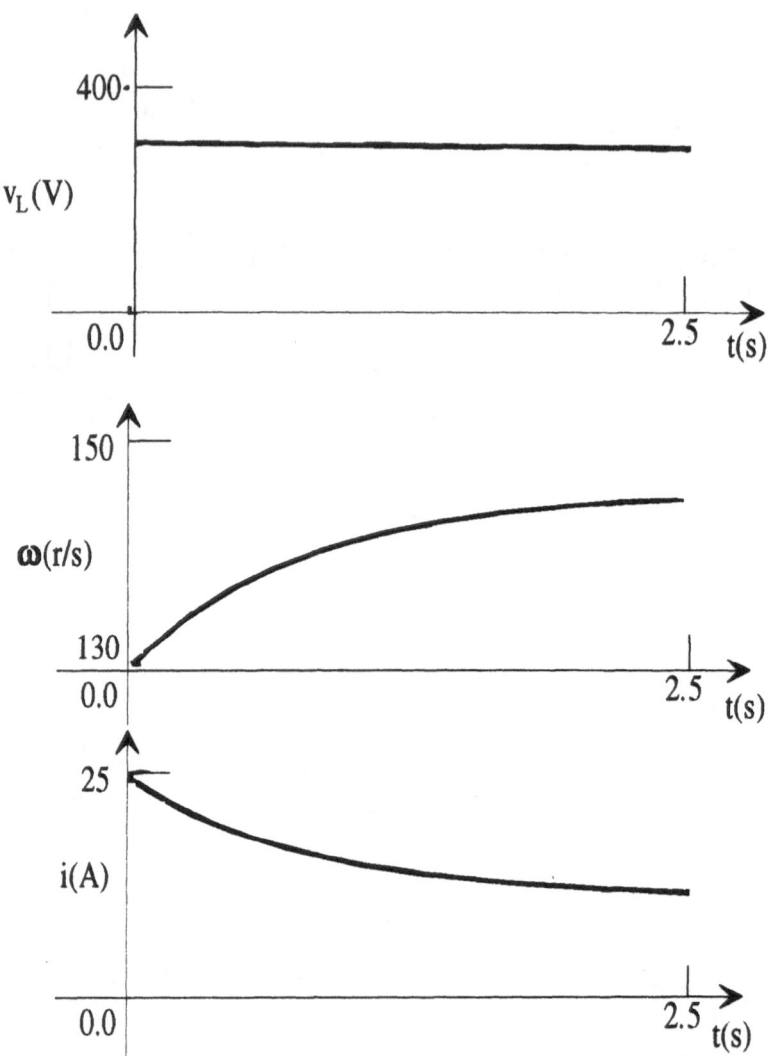

Figure 2.11: Transient response in Example 2.3 for a sudden change in duty cycle from 50% to 75%

2.7 Prediction of limit cycle of van der Pol's oscillator

Let us consider the van der Pol's equation

$$\ddot{f}(t) - \epsilon(1 - f^2(t))\dot{f}(t) + f(t) = 0. \tag{2.44}$$

This equation has been extensively studied in the literature on non-linear systems for its oscillatory behaviour. The standard technique of signal balance or "harmonic balance" has been applied with sinusoidal basis functions for small ϵ, because the limit cycle under this condition is only slightly different from a pure sinusoid. The conventional Fourier series methods of signal balance have therefore been applied with the first sinusoidal term with considerable success [7]. On the other hand, when ϵ is large, the waveform of oscillations tends to assume a trapezoidal form and techniques based on sinusoidal signal balance become enormously complex and expensive since the waveform requires several spectral components for its proper representation.

However, the GHOF spectrum of the waveform of oscillations of van der Pol's equation with large ϵ would have an impressively small size. This leads us to believe that a signal balance technique employing GHOF may be more relevant in such situations. Investigations on the use of Chebyshev polynomials [285] and Walsh functions [12,13,140] in similar nonlinear problems have been reported. In this section we use GHOF as the basis for predicting the amplitude and time-period of oscillations of $f(t)$.

Before we proceed further to formally expand the signals into a series of GHOF we notice that equation (2.44), under the conditions of relaxed oscillations, has $\dot{f}(t)$ and $\ddot{f}(t)$ in the form of sharp and narrow periodic pulses which cannot be economically expanded in terms of any of the systems of orthogonal functions (not even GHOF). In view of this, we integrate (2.44) twice with respect to time for the present study without loss of generality. We then obtain

$$f(t) - \epsilon \int f(t) \, dt + \epsilon/3 \int f^3(t) \, dt + \int\int f(t) \, dt^2 = 0. \tag{2.45}$$

We now expand the expected solution over its half-period T, which itself is an unknown in this case, i.e.,

$$f(t) \approx \mathbf{f}^T \Theta(t), \forall t \in (0, T).\tag{2.46}$$

The terms $\int f(t)\ dt, \int\int f(t)\ dt^2$ and $\int f^3(t)\ dt$ are obtained by the application of operational matrices as follows:

Let

$$g(t) = \int f(t)dt = \mathbf{f}^T \mathbf{E}_g \Theta + \mathbf{g}_0 \Theta = \mathbf{g}^T \Theta,\tag{2.47}$$

$\mathbf{g}_0$ being the GHOF spectrum of the initial condition $g(0)$.

Similarly,

$$h(t) = \int\int f(t)dt^2 = \int g(t)\ dt = \mathbf{g}^T \mathbf{E}_g \Theta + \mathbf{h}_0 \Theta = \mathbf{h}^T \Theta.\tag{2.48}$$

In order to evaluate the integral

$$r(t) = \int f^3(t)\ dt = \mathbf{r}^T \Theta(t),\tag{2.49}$$

one has to first find an approximate representation of $f^3(t)$ in terms of the spectral coefficient vector $\mathbf{f}$ of $f(t)$ and then apply the operational matrix for integration. Let us define the nonlinear operator $\mathbf{N}(\cdot)$ such that

$$f^3(t) \approx \mathbf{N}(\mathbf{f}) \cdot \Theta(t).\tag{2.50}$$

Then,

$$\mathbf{r}^T \Theta(t) = \mathbf{N}(\mathbf{f})\mathbf{E}_g \Theta(t) + \mathbf{r}_0 \Theta(t).\tag{2.51}$$

Here the term $\mathbf{N}(\mathbf{f})$ is a vector of dimension equal to that of $\mathbf{f}$. We may now insert these expansions into equation (2.45):

$$\mathbf{f}^T \Theta(t) - \epsilon\ \mathbf{g}^T \Theta(t) + \epsilon/3\ \mathbf{r}^T \Theta(t) + \mathbf{h}^T \Theta(t) = \boldsymbol{\eta}\ .\tag{2.52}$$

Notice that a vector of errors $\boldsymbol{\eta}$ is allowed on the r.h.s. of (2.52) to account for errors in the approximation. In addition to (2.52) three more conditions imposing periodicity of $g(t), h(t)$ and $r(t)$ are required. Since T is the half-period of oscillations, keeping in view the symmetry of the solution, we have,

$$\left. \begin{array}{rcl} g(T) + g(0) &=& \eta_{\varrho+1} \\ h(T) + h(0) &=& \eta_{\varrho+2} \\ r(T) + r(0) &=& \eta_{\varrho+3} \end{array} \right\} . \tag{2.53}$$

Further, continuity conditions can be imposed at each of the $m - 1$ segment boundaries. We therefore have, a set of nonlinear equations in $\varrho + m + 3$ unknowns. The vector of unknowns is given by

$$\xi^T = [\ \mathbf{f}^T\ |\ T_1\ T_2\ \cdots\ T_m\ |\ g(0)\ h(0)\ r(0)\ ], \tag{2.54}$$

ξ has to be obtained by solving equations (2.52) and (2.53). Several techniques may be adopted for this purpose. We choose the approach B of section 2.5 and use the least squares method by defining

$$\mathcal{J}(\xi) = \sum_i \eta_i^2 . \tag{2.55}$$

$\mathcal{J}$ can now be minimized with respect to the unknowns ξ using any of the standard techniques such as the steepest descent.

Example 2.4. *Solution of van der Pol's equation*

Using the above formulation, approximate solution of (2.45) is now attempted for various values of ϵ. Three cases are considered, all using Legendre polynomials as the continuous basis functions:

(a) $m = 1, r_1 = 2$,

(b) $m = 1, r_1 = 3$ and

(c) $m = 2, r_1 = r_2 = 2$.

$\mathcal{J}$ is minimized by the method of steepest descent. Even though the number of unknowns in the vector $\mathbf{f}$ in the cases (a), (b) and (c) above are 2, 3 and 4 respectively, to accommodate higher order terms generated due to integration, the dimension of $\mathbf{f}$ is taken to be 4, 5 and 8 respectively, with the higher order terms set equal to zero. This ensures that no further loss of accuracy takes place due to the use of the operational matrix for integration in place of actual integration.

The computed results are shown in Tables 2.4– 2.6. Fig. 2.12 shows the limit cycles over the half-period T of oscillations for $\epsilon = 10, 20, 40$ and 100 respectively along with a more accurate result using BPF. In this a large number of terms is used (about 2000) and the segment widths are chosen to be inversely proportional to the derivative of $f(t)$. These plots indicate that as the value of ϵ increases, the waveform of oscillations tends to assume a trapezoidal shape and very little improvement is achieved by considering additional terms in the GHOF approximation.

Table 2.4: Solution of Example 2.4 : Case(a)

ϵ	$f_{1,1}$	$f_{2,1}$	T	$g(0)$	$h(0)$	$r(0)$
10.00	1.670	-0.374	8.058	-6.528	-2.026	-19.782
20.00	1.675	-0.373	16.004	-13.300	-7.999	-39.509
40.00	1.676	-0.372	31.986	-26.745	-31.999	-78.976
100.0	1.676	-0.372	80.000	-67.000	-200.00	-197.40

Table 2.5: Solution of Example 2.4 : Case(b)

ϵ	$f_{1,1}$	$f_{2,1}$	$f_{3,1}$	T	$g(0)$	$h(0)$	$r(0)$
10.00	1.654	-0.404	-0.061	8.390	-6.799	-2.990	-20.109
20.00	1.662	-0.412	-0.020	16.583	-13.436	-13.910	-39.994
40.00	1.681	-0.385	-0.016	32.538	-27.361	-55.961	-80.031
100.0	1.693	-0.364	-0.023	80.269	-69.658	-349.99	-200.02

Table 2.6: Solution of Example 2.4 : Case(c)

ϵ	$f_{1,1}$	$f_{2,1}$	$f_{1,2}$	$f_{2,2}$ $g(0)$	T_1 $h(0)$	T_2 $r(0)$
10.00	1.680	-0.364	-0.604	-1.513	8.121	0.303
				-6.726	-3.461	-19.955
20.00	1.650	-0.415	-0.485	-1.789	16.957	0.174
				-13.644	-13.939	-39.787
40.00	1.648	-0.429	-0.453	-1.971	33.772	0.429
				-27.260	-55.925	-79.657
100.0	1.684	-0.367	-0.468	-1.525	80.263	1.232
				-69.608	-349.99	-200.04

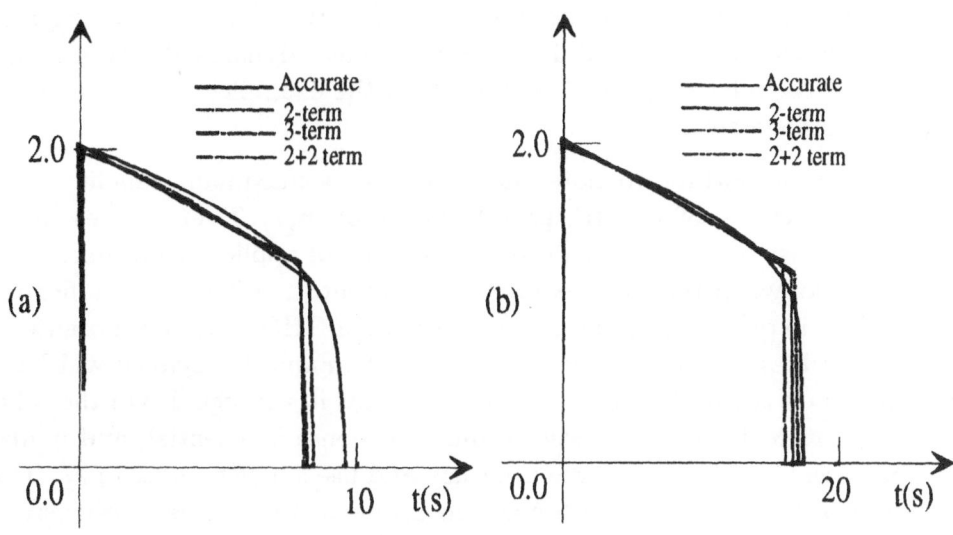

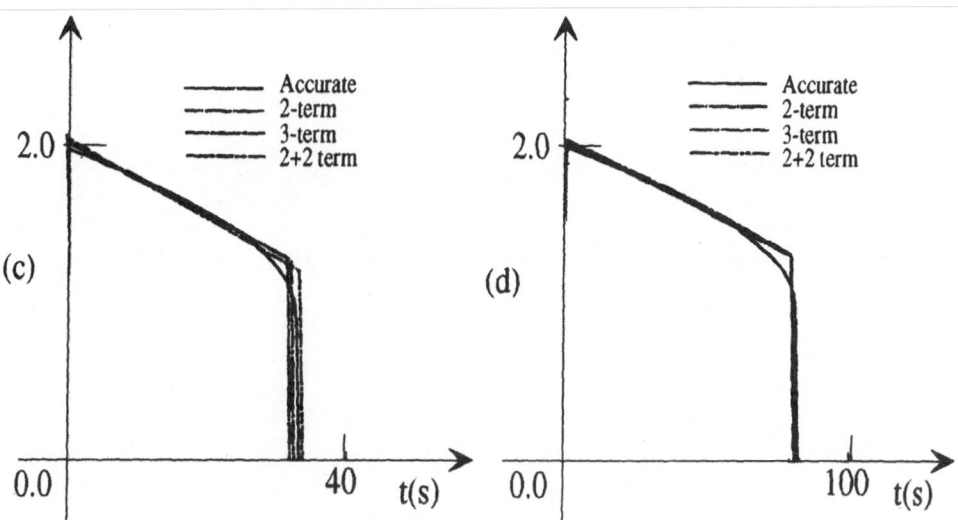

Figure 2.12: Solution of Example 2.4 for (a) $\epsilon = 10$, (b) $\epsilon = 20$, (c) $\epsilon = 40$ and (d) $\epsilon = 100$

2.8   Remarks

In this chapter the proposed system of GHOF has been successfully applied to the analysis of linear and nonlinear dynamical systems. The size of the related spectral vectors in all the examples turned out to be impressively small.

It is important to note that, to extract maximum benefit out of the framework of GHOF, proper choice of m, r_j, T_j etc. is essential. They must be chosen with the specific type of application in mind. For example, when the time required for computation is of no significance, such as in off-line applications, the system of GHOF may be chosen with large values of r_j to improve accuracy, with m and the segment widths T_j chosen on the basis of discontinuities, if any, in the signal. On the other hand, in real-time applications, quick response is essential, and it may be necessary to compromise accuracy and use smaller values of r_j. This kind of trade-off has to be made in problems of parameter estimation and adaptive control as will be seen in the next two chapters.

Chapter 3

Identification of Continuous-time Systems

3.1 Survey of literature in the field

One of the important advantages of characterization of signals in a system by orthogonal functions is that the continuous-time (CT) model can be directly handled. When one is interested in finding out only the response of the system, it may not seem very advantageous. But, when we are concerned with the identification of such systems, this has important consequences. In fact, identification of CT model parameters itself is important for several reasons. Some of these have been pointed out by Edmunds [69] and Unbehauen and Rao [282] in the context of parameter estimation and by Gawthrop [78,77] in the context of self-tuning control. We briefly discuss them here.

i) Our understanding of systems and control has been largely through the CT models. The properties of naturally continuous systems are easier to interpret through CT models than their discrete-time (DT) counterparts.

ii) Sometimes indirect approaches to CT model identification are used in which a suitable DT model is first identified, and the CT model is obtained through an appropriate transformation. These approaches suffer from many difficulties because the coefficients of the DT transfer function become numerically ill-conditioned [69] and many of the DT to CT transformations are also associated with similar problems [249]. These effects are more prominent when the sampling rate is high.

iii) The relative order information (number of poles − number of zeros) is lost in sampling. The additional zeros which arise out of discretization lie outside the unit circle for relative orders greater than two [5]. Furthermore, when the original CT system has a time-delay, the resulting DT approximation may turn out to be non-minimum phase, if the time-delay is not an integral multiple of the sampling period [77], even if the original CT model is minimum phase.

iv) A frequent further use of an identified model is in self-tuning control. The properties of a DT control depend on how accurately the model parameters are represented. For example, Edmunds has shown [69] that an error of 1 part in 10^7 in the coefficient of z^3 in the z-transfer function (for a 6-th order CT system sampled at 0.1s) could make a 50% change in the steady state gain of the model. The situation worsens with increasing sampling rates. Since none of the existing parameter estimation schemes is likely to estimate parameters with such a high precision, when the estimated model is used in the computation of controller parameters, the solution may lead to completely erroneous results and to numerical ill-conditioning [61,62].

These problems can be avoided if the continuous-time model is directly identified. Young surveyed the various approaches to parameter estimation of continuous-time models in 1981 [308]. Subsequently some books have appeared on this topic [216,247,282]. Orthogonal functions were introduced to the problem of parameter estimation of linear time-invariant continuous time models in 1975 [224] with Walsh functions as basis. Use of other basis functions followed in subsequent years [267,145,188,18,194,198]. Some alternative approaches to such direct estimation were proposed at the same time. Saha and Rao [241] used the so-called Poisson moment functionals. Recently techniques based on numerical integration and digital filtering have been reported [236,237,238,303].

The activity in the field during the period following Young's survey has been summarized in Table 3.1. The details of the abbreviations are given in "List of Abbreviations". For completeness, indirect approaches and those based on Poisson moment functionals, numerical integration and Taylor series are also included in this table.

In this chapter we first solve the parameter estimation problem in linear CT models using the GHOF. This formulation is then recursified to suit real-time applications. For this purpose, a recursive formula for computing multiple integrals of a signal characterized by GHOF spectra

Table 3.1: Literature map for identification of continuous-time systems

Tech-	Types of systems							
nique	LT	TD	TV	NL	MV	DT	SC	Others
BPF	[50,129, 145,267, 300,202, 201]	[127, 260]	[108, 125, 300]	[42,49, 124, 301]	[127, 15, 182]			[133, 266, 222, 183]
WF	[9,8,10,67, 68,188,192, 219,224, 227,235, 280,286]	[11, 75, 188, 225]		[44, 132, 228]	[193, 226]	[168]		[94, 134]
GBPF	[287,298]	[295]	[293]	[298]				
PCBF			[96]		[97, 96]			
OP	[310]		[270]	[138]	[102]			[57, 301]
LAP	[16,63,108, 113,142]	[16, 143]		[215, 309]		[199]		[115, 256]
LEP	[18,23,174, 196]	[23, 104, 146, 255]	[54, 110]	[54,93, 105, 106, 252, 255]	[111]	[122]	[253]	[91, 19, 302]
CHP	[194,158]	[88]	[56, 55, 178, 261]	[56,58, 157, 254]			[52]	[89, 121, 160]
JAP	[159]						[51]	
HEP	[198]	[210]						
GOP	[299]		[26]	[147]				
SCF	[59,175]							
NI	[236,238, 237,239, 303]	[212, 211]			[31]			[302]
DT	[213,214, 264,283, 262,139, 269,69]	[1]		[76]	[265, 263]			[161, 214, 311]
PMF,MF	[74,223, 241,243, 245,240]	[244]	[242]	[209, 207, 208]	[246]			[74, 240]
TS	[179]	[33]	[180, 270]	[120]		[120]	[60]	
GHOF	[155,206]							

is derived. Utilizing this, a recursive least squares parameter estimation algorithm based on GHOF is proposed and applied to the problem of parameter estimation in a converter driven DC motor system. Then a generalized least squares parameter estimation algorithm employing block pulse functions is presented and its performance is studied by Monte Carlo simulations. Finally, an algorithm for simultaneous estimation of parameters and state of a continuous-time system is proposed and numerical results showing its performance are presented.

3.2   Formulation of the problem in terms of GHOF spectra

Consider a single-input single-output (SISO) CT linear time-invariant system modelled by the differential equation

$$\sum_{l=0}^{n} a_l \frac{d^l y}{dt^l} = \sum_{l=0}^{n_b} b_l \frac{d^l u}{dt^l}, \tag{3.1}$$

where $u(t)$ and $y(t)$ are the input and output signals of the system respectively. Integrating (3.1) n-times with respect to t,

$$\sum_{l=0}^{n} a_l\, y_{(n-l)}(t) = \sum_{l=0}^{n_b} b_l\, u_{(n-l)}(t) + \sum_{l=0}^{n-1} \gamma_l\, \iota_{(l)}(t) \tag{3.2}$$

where, $y_{(l)}(t), u_{(l)}(t)$ and $\iota_{(l)}(t)$ are the l-th integrals of $y(t)$, $u(t)$ and the unit step function $\iota(t)$ between the limits $(0, t)$ respectively. $\{\gamma_l\}$ is the set of initial condition terms arising out of integration of (3.1).

Expanding $u(t)$, $y(t)$ and $\iota(t)$ in terms of their GHOF spectra,

$$\begin{aligned} y(t) &= \mathbf{y}^T \mathbf{\Theta}(t) \\ u(t) &= \mathbf{u}^T \mathbf{\Theta}(t) \\ \iota(t) &= \iota^T \mathbf{\Theta}(t) \end{aligned} \left.\right\} \tag{3.3}$$

and substituting in (3.2),

$$\sum_{l=0}^{n} a_l\, \mathbf{y}^T \mathbf{E}_g^{n-l}\, \mathbf{\Theta}(t) = \sum_{l=0}^{n_b} b_l\, \mathbf{u}^T \mathbf{E}_g^{n-l}\, \mathbf{\Theta}(t) + \sum_{l=0}^{n-1} \gamma_l\, \iota^T \mathbf{E}_g^{l}\, \mathbf{\Theta}(t). \tag{3.4}$$

Setting $a_n = 1$ without loss of generality, (3.4) can be written as

$$\mathbf{y} = -\sum_{l=0}^{n-1} a_l \, \mathbf{E}_g^{T^l} \mathbf{y} + \sum_{l=0}^{n_b} b_l \, \mathbf{E}_g^{T^l} \mathbf{u} + \sum_{l=0}^{n-1} \gamma_l \mathbf{E}_g^{T^l} \iota, \tag{3.5}$$

or,

$$\mathbf{y} = \mathbf{M}\,\mathbf{p}, \tag{3.6}$$

where,

$$\mathbf{M} \;=\; [\,-\mathbf{E}_g^{T^n}\mathbf{y}\,|-\mathbf{E}_g^{T^{n-1}}\mathbf{y}\,|\cdots|-\mathbf{E}_g^{T}\mathbf{y}\,|\,\mathbf{E}_g^{T^n}\mathbf{u}\,|\,\mathbf{E}_g^{T^{n-1}}\mathbf{u}\,|$$
$$|\cdots|\,\mathbf{E}_g^{T^{n-n_b}}\mathbf{u}\,|\,\iota\,|\,\mathbf{E}_g^{T}\iota\,|\cdots|\,\mathbf{E}_g^{T^{n-1}}\iota\,]$$

and

$$\mathbf{p}^T = [\,a_0 \, a_1 \, \ldots \, a_{n-1} \,|\, b_0 \, b_1 \, \ldots \, b_{n_b} \,|\, \gamma_0 \, \gamma_1 \, \ldots \, \gamma_{n-1}\,].$$

The problem is to estimate the parameter vector $\mathbf{p}$ based on the knowledge of $\mathbf{y}$ and $\mathbf{M}$ in which the signals are characterized by GHOF and operated upon by $\mathbf{E}_g$. The estimation can now be done by many of the existing methods. To illustrate the one based on the least squares (LS) approach, we consider the error vector

$$\mathbf{e} = \mathbf{y} - \mathbf{M}\,\mathbf{p} \tag{3.7}$$

and define the cost functional

$$\mathcal{J} = \mathbf{e}^T \mathbf{e} = (\mathbf{y} - \mathbf{Mp})^{\mathbf{T}} \, (\mathbf{y} - \mathbf{Mp}). \tag{3.8}$$

Minimizing $\mathcal{J}$ with respect to the unknown parameter vector $\mathbf{p}$, we obtain the LS estimate $\hat{\mathbf{p}}$ of $\mathbf{p}$ as:

$$\hat{\mathbf{p}} = (\mathbf{M}^T\mathbf{M})^{-1}\mathbf{M}^T\mathbf{y}. \tag{3.9}$$

Several modifications of the LS scheme are possible. These include the introduction of a weighting matrix in the definition of $\mathcal{J}$, the use of the generalized least squares (GLS) and instrumental variable methods etc. The latter approaches are used when the input-output data are heavily corrupted by noise and the simple LS scheme yields biased parameter estimates.

3.3 Recursive computation of multiple integrals of a signal

The computation of $\hat{\mathbf{p}}$ using (3.9) involves a computational complexity of the order of $n_p^3 + (\varrho+1)n_p^2 + \varrho n_p$ multiplications where $n_p = 2n + n_b + 1$ is the number of parameters to be estimated. In real-time applications, where the data available has to be processed before the next set of measurements arrives, it is more appropriate to compute the estimates using a recursive relation which updates them over successive intervals of time. Such recursive algorithms have been extensively used in the discrete-time context [162]. Here we are interested in deriving such a recursive algorithm for the GHOF scheme.

We notice that in order to derive the recursive algorithm it is necessary to obtain the various rows of the matrix $\mathbf{M}$ recursively. This involves recursive computation of the multiple integrals of the signals $y(t), u(t)$ and $\iota(t)$ respectively. Such a scheme is presented here.

One way of recursifying the above computation is to describe the multiple integrator in state space in a suitable canonical form in which the state variables represent the integrals of various orders. Then the SSMT solution of section 2.4 may be employed to compute these integrals recursively. However, this method is computationally expensive since it is necessary to perform a matrix inversion (if either the time segments T_j or the number of CBF components r_j are unequal) and a matrix vector multiplication at each recursive step. A simpler approach is to recursify the computation of integrals making use of the special structure of $\mathbf{E}_g$ in which the elements $h_{l,j,i}$ in $\mathbf{H}_{l,j}$ are, in fact, independent of j (section 2.2).

Let us consider

$$g(t) = \int_0^t f(t) \, dt, \tag{3.10}$$

which implies that

$$\mathbf{g} = \mathbf{E}_g \, \mathbf{f}, \tag{3.11}$$

where $\mathbf{f}$ and $\mathbf{g}$ are the GHOF spectral vectors of $f(t)$ and $g(t)$ respectively. If the j-th subvector of $\mathbf{f}$ corresponding to the segment Ω_j is defined as

$$\mathbf{f}_j^T = [\ f_{1,j}\ f_{2,j}\ \cdots\ f_{r_j,j}\ ]$$

and similarly,

$$\mathbf{g}_j^T = [\ g_{1,j}\ g_{2,j}\ \cdots\ g_{r_j,j}\ ],$$

then, in view of the structure of $\mathbf{E}_g$ in (2.6),

$$\mathbf{g}_j = \mathbf{E}_j^T\, \mathbf{f}_j + \sum_{l=1}^{j-1} \mathbf{H}_{l,j}^T\, \mathbf{f}_l. \tag{3.12}$$

From (2.7)

$$\mathbf{H}_{l,j}^T\, \mathbf{f}_l = \sum_{l=1}^{r_l} h_{l,j,i}\, f_{i,l}\, \nu_j, \tag{3.13}$$

where, ν_j is defined by (2.26).

Therefore,

$$\mathbf{g}_j = \mathbf{E}_j^T\, \mathbf{f}_j + \sum_{l=1}^{j-1} \sum_{i=1}^{r_l} h_{l,j,i}\, f_{i,l}\, \nu_j. \tag{3.14}$$

Similarly,

$$\mathbf{g}_{j+1} = \mathbf{E}_{j+1}^T\, \mathbf{f}_{j+1} + \sum_{l=1}^{j} \sum_{i=1}^{r_l} h_{l,j+1,i}\, f_{i,l}\, \nu_{j+1}. \tag{3.15}$$

Denoting,

$$\zeta_j = \sum_{l=1}^{j-1} \sum_{i=1}^{r_l} h_{l,j,i}\, f_{i,l}, \tag{3.16}$$

$$\zeta_{j+1} = \zeta_j + \sum_{i=1}^{r_j} h_{j,j,i}\, f_{i,j}, \tag{3.17}$$

since, $h_{l,j+1,i} = h_{l,j,i}$.

Therefore, the recursive relation for computation of integrals is given by

$$\mathbf{g}_{j+1} = \mathbf{E}_{j+1}^{T}\mathbf{f}_{j+1} + \zeta_{j+1}\,\boldsymbol{\nu}_{j+1} \tag{3.18}$$

where ζ_{j+1} can be calculated recursively from (3.17).

Multiple integrals can thus be computed by applying (3.18) repeatedly. This recursification is in terms of a block of size r_j, the dimension of the GHOF spectral vector over the segment Ω_j. With these elements, each spectral component in the vector may be separated out to form an r_j-set of independent linear equations which may then be processed successively through a recursive parameter estimation algorithm.

Using the above formulation, equation (3.6) may be modified as

$$\mathbf{y}_j = \boldsymbol{\Phi}_j\,\mathbf{p}, \ \forall\, j \in \mathcal{I}_m, \tag{3.19}$$

where,

$$\mathbf{y}_j^T = [\ y_{1,j}\ y_{2,j}\ \cdots\ y_{r_j,j}\ ],$$

and

$$\boldsymbol{\Phi}_j = [-\mathbf{y}_{n,j}| - \mathbf{y}_{n-1,j}| \cdots |\mathbf{y}_{1,j}|\mathbf{u}_{n,j}|\mathbf{u}_{n-1,j}| \cdots |\mathbf{u}_{n-n_b,j}|$$
$$\iota_{1,j}|\iota_{2,j}| \cdots |\iota_{n-1,j}]$$

in which $\mathbf{y}_{l,j}$, $\mathbf{u}_{l,j}$ and $\iota_{l,j}$ are defined to be the GHOF spectra of $y_{(l)}(t)$, $u_{(l)}(t)$ and $\iota_{(l)}(t)$ respectively over the segment Ω_j.

3.4 Recursive least squares (LS) parameter estimation algorithm employing GHOF

We now employ the recursive least squares parameter estimation algorithm within the GHOF framework. This is a modified version of the discrete-time algorithm given in [162], given by the following set of equations:

$$\left. \begin{aligned}
k &= 0 \\
\text{For } j &= 1, 2, \ldots, m \text{ do} \\
\text{For } i &= 1, 2, \ldots, r_j \text{ do} \\
k &= k+1 \\
\hat{e}(k+1) &= y_{i,j} - \phi_{i,j}^T \, \hat{\mathbf{p}}(k) \\
\mathbf{q}(k+1) &= \mathbf{P}(k) \, \phi_{i,j} \, [\rho(k+1) + \phi_{i,j}^T \, \mathbf{P}(k) \, \phi_{i,j}]^{-1} \\
\hat{\mathbf{p}}(k) &= \hat{\mathbf{p}}(k) + \mathbf{q}(k+1) \, \hat{e}(k+1) \\
\mathbf{P}(k+1) &= [\mathbf{P}(k) - \mathbf{q}(k+1) \, \phi_{i,j}^T \, \mathbf{P}(k)] / \rho(k+1)
\end{aligned} \right\} , \quad (3.20)$$

where,

$$
\begin{aligned}
\hat{e} &= \text{Equation error} \\
\hat{\mathbf{p}} &= \text{Estimate of the parameter vector} \\
\mathbf{P} &= \text{Covariance matrix} \\
\phi_{i,j}^T &= i\text{-th row of } \mathbf{\Phi}_j \\
\mathbf{q} &= \text{Kalman gain vector} \\
\rho &= \text{Forgetting factor, normally chosen between } (0.95, 1.0).
\end{aligned}
$$

To start the algorithm, $\hat{\mathbf{p}}$ has to be initialized with suitable estimates of the parameters and $\mathbf{P}(1) = \sigma \mathbf{I}$ where σ is a large positive value. For stability of the algorithm $\mathbf{P}(k)$ should be positive definite for all k and several schemes to ensure this are employed in practice [163,86].

3.5 Parameter estimation in a converter driven DC motor system

Let us again consider the system of Example 2.2. The various DC motor parameters such as R, L, K_a, K_T etc. along with the load torque T_L are assumed to be unknown. These are to be estimated based on the measurements of the terminal voltage $u_1(t)$, motor current $x_1(t)$ and the speed $x_2(t)$.

From (2.39),

$$\dot{\mathbf{x}}_1 = -\frac{R}{L} x_1 - \frac{K_a}{L} x_2 + \frac{1}{L} u_1 \qquad (3.21)$$

$$\dot{\mathbf{x}}_2 = \frac{K_T}{J} x_1 - \frac{1}{J} u_2. \qquad (3.22)$$

Integrating with respect to time,

$$x_1 = -\frac{R}{L}\int_0^t x_1(t)\,dt - \frac{K_a}{L}\int_0^t x_2(t)\,dt + \frac{1}{L}\int_0^t u_1(t)\,dt + x_1(0). \quad (3.23)$$

We expand x_1, x_2, u_1 and u_2 in terms of their GHOF spectra,

$$x_1(t) \approx \mathbf{x}_1\Theta(t),\quad x_2(t) \approx \mathbf{x}_2\Theta(t),\quad u_1(t) \approx \mathbf{u}_1\Theta(t),\quad u_2(t) \approx \mathbf{u}_2\Theta(t),$$

and define

$$x_1(0) = \mathbf{x}_{1,0}^T\,\Theta(t), \text{ where } \mathbf{x}_{1,0} = x_1(0)\,\boldsymbol{\nu}$$

with $\boldsymbol{\nu}$ defined by (2.25).

Then,

$$\mathbf{x}_1 = -\frac{R}{L}\mathbf{E}_g^T\,\mathbf{x}_1 - \frac{K_a}{L}\mathbf{E}_g^T\,\mathbf{x}_2 + \frac{1}{L}\mathbf{E}_g^T\,\mathbf{u}_1 + \mathbf{x}_{1,0}, \qquad (3.24)$$

or,

$$\mathbf{x}_1 = [\ -\mathbf{E}_g^T\mathbf{x}_1|-\mathbf{E}_g^T\mathbf{x}_2|\mathbf{E}_g^T\mathbf{u}_1|\boldsymbol{\nu}\ ]\ [\ \tfrac{R}{L}\ \tfrac{K_a}{L}\ \tfrac{1}{L}\ x_1(0)\ ]^T. \qquad (3.25)$$

Similarly, from (3.22),

$$\mathbf{x}_2 = [\ \mathbf{E}_g^T\mathbf{x}_1|-\mathbf{E}_g^T\mathbf{u}_2|\boldsymbol{\nu}\ ][\ \tfrac{K_T}{J}\ \tfrac{1}{J}\ x_2(0)\ ]^T. \qquad (3.26)$$

Equations (3.25) and (3.26) are derived from (2.39) which is valid during the *on*-period of the converter. From (2.40), which is valid during the *off*-period, we obtain,

$$\mathbf{x}_2 = [\ \mathbf{E}_g^T\mathbf{x}|\ \boldsymbol{\nu}\ ][\ \tfrac{1}{J}\ x_2(0)\ ]^T. \qquad (3.27)$$

Note that since $x_1(t) = 0$ in this case, only the equation relating the speed and the load torque is relevant.

Making use of the recursive relation (3.18), (3.25) becomes

$$\mathbf{x}_{1,j} = [\ -\psi_{1,j}\ |\ -\psi_{2,j}\ |\ \psi_{3,j}|\ \nu_j\ ][\ p_1\ p_2\ p_3\ p_4\ ]^T, \qquad (3.28)$$

where $\psi_{1,j}, \psi_{2,j}$ and $\psi_{3,j}$ are the GHOF spectra of the integrals of the signals $x_1(t), x_2(t)$ and $u_1(t)$ over Ω_j respectively, and $p_1 = \frac{R}{L}, p_2 = \frac{K_a}{L}, p_3 = \frac{1}{L}$ and $p_4 = x_1(0)$.

Similarly, (3.26) yields,

$$\mathbf{x}_{2,j} = [\, \psi_{1,j} \, | \, -\psi'_{4,j} \, | \, \nu_j \,][\, p_5 \; p'_6 \; p_7 \,]^T, \tag{3.29}$$

where $\psi'_{4,j}$ is the GHOF spectral vector of the integral of $u_2(t)$ over Ω_j and $p_5 = \frac{K_T}{J}, p'_6 = \frac{1}{J}$ and $p_7 = x_2(0)$.

If the load torque is also treated as an unknown [123], (3.29) may be modified to

$$\mathbf{x}_{2,j} = [\, \psi_{1,j} \, | \, -\psi_{4,j} \, | \, \nu_j \,][\, p_5 \; p_6 \; p_7 \,]^T, \tag{3.30}$$

where $\psi_{4,j}$ is the GHOF spectral vector of a unit ramp function over Ω_j and $p_6 = \frac{T_L}{J}$.

Similarly, (3.27) leads to

$$\mathbf{x}_{2,j} = [\, -\psi_{4,j} \, | \, \nu_j \,][\, p_6 \; p_7 \,]^T. \tag{3.31}$$

Equations (3.28)–(3.31) are now ready for recursive computation via (3.20). Noting the similarity in the parameter vectors in (3.30) and (3.31), we can combine them by padding (3.31) with a zero term. In fact, (3.28) can also be combined with (3.30) and (3.31) using suitable padding, but at the cost of considerable increase in computational complexity due to the loss of natural decentralized nature of the estimation equations. Therefore the forms of the $\phi_{i,j}$ vectors corresponding to the three equations (3.28), (3.30) and (3.31) are given by

$$\phi_{i,j}^T = i\text{-th row of } [\, -\psi_{1,j} \, | \, -\psi_{2,j} \, | \, \psi_{3,j} \, | \, \nu_j \,],$$

$$\phi_{i,j}^T = i\text{-th row of } [\, \psi_{1,j} \, | \, -\psi_{4,j} \, | \, \nu_j \,] \text{ and}$$

$$\phi_{i,j}^T = i\text{-th row of } [\, \mathbf{0} \, | \, -\psi_{4,j} \, | \, \nu_j \,].$$

Example 3.1. Parameter estimation in a DC motor system

The system of Example 2.2 is now considered for parameter estimation. The input-output data is generated by the same simulation scheme and the GHOF spectra of these signals are directly obtained. When normal operating records are used, these have to be calculated using the inner product formula (1.9). Transient data as shown in Fig. 2.7 has been used upto 100 cycles (the definition of the term *cycle* also remains the same as before). For the purpose of parameter estimation, the voltage and current waveforms are approximated by two Legendre terms in each cycle while the speed waveform is represented by a single Legendre component. The higher order terms are ignored and the equations arising out of them are not processed by the parameter estimation algorithm. Similarly, the *off*-periods in each cycle are seen to contribute very little in view of their small duration. Therefore, the data obtained during this period may also be ignored. However, the computed integral terms must take into account the values of the signals during this period. For example, the back e.m.f. which comes across the motor terminals during this period, must be used in the calculation of the integral of the input terminal voltage $v_t(t) = u_1(t)$.

Figure 3.1 shows the pattern of parameter convergence for 100 cycles of input-output data. Initially the parameters are assumed to be zero and σ is chosen to be 10^4. The parameters have converged well and the estimated values of the motor parameters are

$$\hat{R} = 0.097\Omega, \ \hat{L} = 0.0055\text{H}, \ \hat{K}_a = 2.48\text{V-s/rad};$$

$$\hat{J} = 1.775\text{Kg-m}^2, \ \hat{K}_T = 2.58\text{N-m/A} \text{ when } T_L \text{ is assumed known.}$$

The above values are very close to the true values given in Example 2.2.

3.6 Parameter estimation using generalized least squares (GLS) scheme

It is well-known in the context of discrete-time parameter estimation that the ordinary least squares scheme gives biased parameter estimates

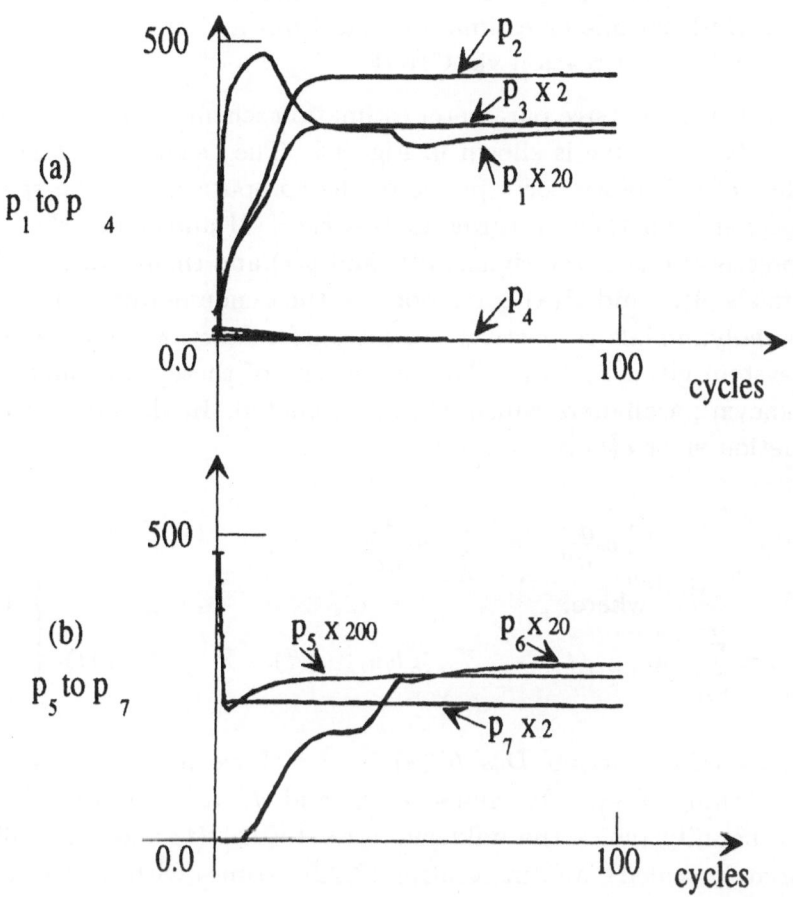

Figure 3.1: Pattern of parameter convergence in Example 3.1

when the input-output data is corrupted by noise. The orthogonal-functions-based schemes are immune to additive high-frequency zero-mean white noise to some extent, since the process of finding the spectra of signals is a kind of low-pass filtering. But since the minimization of an equation error (which does not become white even when the additive noise term is so), is involved, biased parameters are likely to be obtained in noisy situations. It is therefore necessary to use algorithms with better noise rejection property. Here we consider the generalized least squares (GLS) parameter estimation algorithm and employ it for CT-model parameter estimation via GHOF.

The general recursive parameter estimation scheme using CT models using the GLS scheme is shown in Fig. 3.2. The "signal processing or characterization" block corresponds to the computation of the spectra of the signals and their integrals in this case. U and Y are the spectral representations of the signals $u(t)$ and $y(t)$ and their integrals. The polynomials $A(s)$ and $B(s)$ correspond to the denominator and the numerator polynomials respectively in a transfer function representation of the system given by (3.1). The coefficients of these polynomials are the unknown parameters which will be estimated. In the GLS scheme, the equation error $e(t)$ is given by,

$$
\left.
\begin{aligned}
\sum_{l=0}^{n} c_l \frac{d^l e}{dt^l} \; &= \; \sum_{l=0}^{n} d_l \frac{d^l v}{dt^l} \\
\text{where,} & \\
v = \sum_{l=0}^{n} a_l y_{(n-l)}(t) \; &- \; \textstyle\sum_{l=0}^{n_b} b_l u_{(n-l)}(t) - \sum_{l=0}^{n-1} \gamma_l u_{(l)}(t)
\end{aligned}
\right\} .(3.32)
$$

The transfer function $D(s)/C(s)$ in the above may be viewed as a pre-whitening filter. Assuming a_n, c_n and d_n to be unity without loss of generality (since the unknown error terms $e(t)$ and $v(t)$ will be appropriately scaled), and integrating (3.22) n-times with respect to t,

$$
\left.
\begin{aligned}
e(t) + \sum_{l=0}^{n-1} c_l e_{(n-l)}(t) = y(t) \; &+ \; \sum_{l=0}^{n-1} a_l y_{(n-l)}(t) \\
-\sum_{l=0}^{n_b} b_l u_{(n-l)}(t) - \sum_{l=0}^{n-1} \gamma_l u_{(l)}(t) \; &+ \; \sum_{l=0}^{n-1} d_l v_{(n-l)}(t)
\end{aligned}
\right\}, \quad (3.33)
$$

where $e_{(l)}(t)$ denotes the l-th integral of $e(t)$ within the limits $(0, t)$. Equation (3.33) can be compactly written as

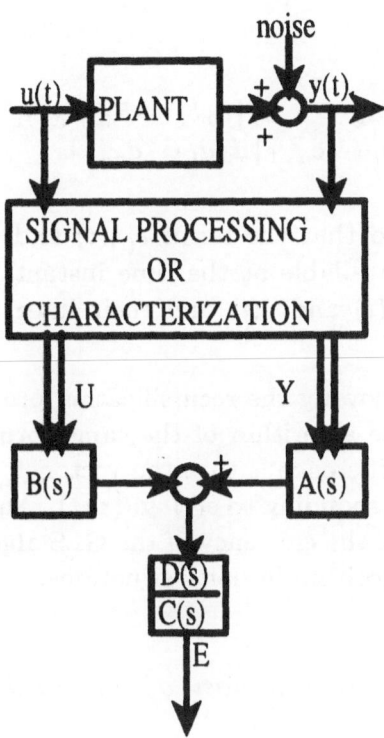

Figure 3.2: GLS scheme for parameter estimation of CT systems

$$e(t) = y(t) - \phi^T(t)\,\mathbf{p} \tag{3.34}$$

where,

$$\begin{aligned}
\phi^T(t) =\ & [\,-y_{(n)}(t)\,|\,\cdots\,|-y_{(1)}(t)\,|\,u_{(n)}(t)\,|\,\cdots\,|\,u_{(n-n_b)}(t)\,| \\
& \iota(t)\,|\,\iota_{(1)}(t)\,|\,\cdots\,|\,\iota_{(n-1)}(t)\,|\,e_{(n)}(t)\,|\,\cdots\,|\,e_{(1)}(t)\,| \\
& -v_{(n)}(t)\,|\,\cdots\,|-v_{(1)}(t)\,]
\end{aligned}$$

and

$$\begin{aligned}
\mathbf{p}^T =\ & [\,a_0\,a_1\cdots a_{n-1}\,|\,b_0\,b_1\cdots b_{n_b}\,|\,\gamma_0\,\gamma_1\cdots\gamma_{n-1}\,| \\
& c_0\,c_1\cdots c_{n-1}\,|\,d_0\,d_1\cdots d_{n-1}\,].
\end{aligned}$$

It may be noted that the terms $e_{(l)}(t)$ and $v_{(l)}(t)$ in $\phi(t)$ will not be measurable or available at the time instant t. Therefore, only the estimated values $\hat{e}(t)$ and $\hat{v}(t)$ and their integrals can be inserted in their place.

From (3.34), following the recursification procedure outlined in section 3.3, a recursive algorithm of the same form as (3.20) can be obtained. The only difference is that the vector $\phi_{i,j}$ will now contain additional terms corresponding to $\hat{e}(t)$ and $\hat{v}(t)$. An example is considered here to demonstrate the efficiency of the GLS algorithm as compared to the ordinary LS algorithm in noisy situations.

Example 3.2. Comparison of LS and GLS schemes in presence of noise

Let us consider a second order linear time-invariant system modelled by

$$\frac{d^2y}{dt^2} + 2\frac{dy}{dt} + y(t) = u(t)$$

with $u(t)$ as a PRBS signal of period 127 and amplitude ±1. The system is simulated using BPF with a time-width of 0.1s and different levels of zero-mean Gaussian noise are added to the BPF coefficients of $y(t)$. The signal to noise ratio (SNR) is defined by

$$\text{SNR} = \frac{\text{Standard deviation of signal}}{\text{Standard deviation of noise}}.$$

1000 points of data spanning a length of 100s are considered. The input signal, the noise-free output signal and the output signal mixed with noise (SNR=3.3) are shown in Fig. 3.3(a)–(c). The other conditions are as follows:

i) BPF are used for signal characterization. Further, the BPF coefficients of outputs, mixed with noise, are directly processed by the parameter estimation algorithms. The inherent filtering property of GHOF therefore does not come into picture in this example. From points of view of susceptibility to noise, this situation therefore corresponds to the worst case.

ii) The GLS algorithm is used with $D(s) = s^n$. The corresponding $e(t)$ terms in the measurement vector at the $(j+1)$-th time interval are substituted by their values at the j-th instant.

Table 3.2 shows the estimates of the parameters for different SNR along with the standard deviation of the parameters from the average value for 20 Monte Carlo simulation runs in each case. The pattern of parameter convergence for the noise-free case and at an SNR of 3.3 is shown for the LS and the GLS schemes in Figures 3.4(a)–(d). It is clear from these plots and Table 3.2 that the GLS scheme is considerably superior to the ordinary LS scheme in presence of noise. In fact, the LS algorithm cannot be used at all in presence of noise, as indicated by the very large standard deviations caused by the lack of parameter convergence.

3.7 Simultaneous state and parameter estimation of SISO systems

In section 3.3 we have recursively estimated the parameters of a continuous-time system along with some initial condition terms $\{\gamma_i\}$, which seems to be, on the surface, an additional burden on the algorithm. However, this unavoidable burden can be viewed as a bonus in the event of simultaneous parameter and state estimation. The estimated system parameters are in the first and second subvectors of $\hat{p}$. The third subvector in $\hat{p}$ may be related to the initial system state in a canonical form. The subvectors of $\hat{p}$ along with the input-output signals can thus be used to directly estimate the system state.

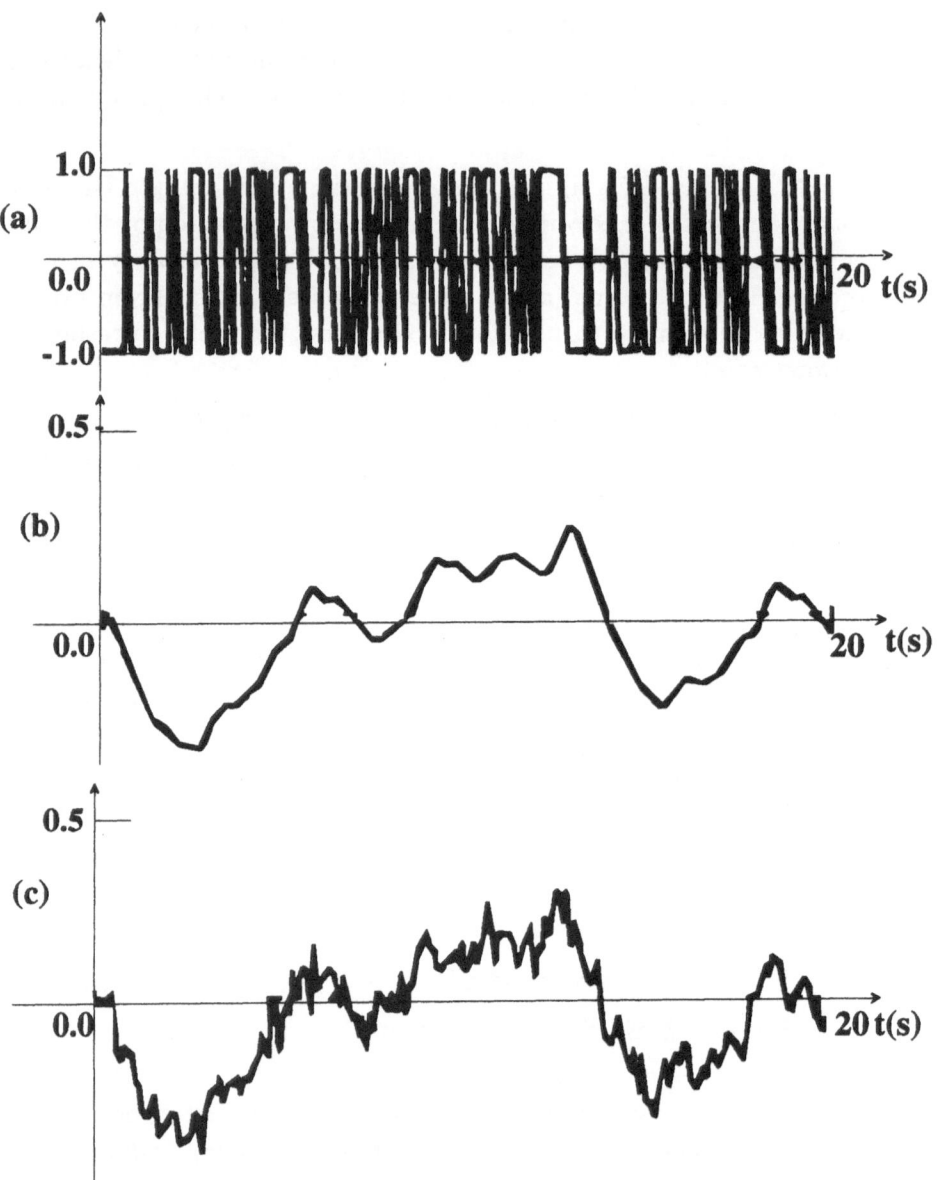

Figure 3.3: Input-output data for parameter estimation in Example 3.2. (a) Input signal; (b) Noise-free output signal and (c) Output signal for SNR=3.3

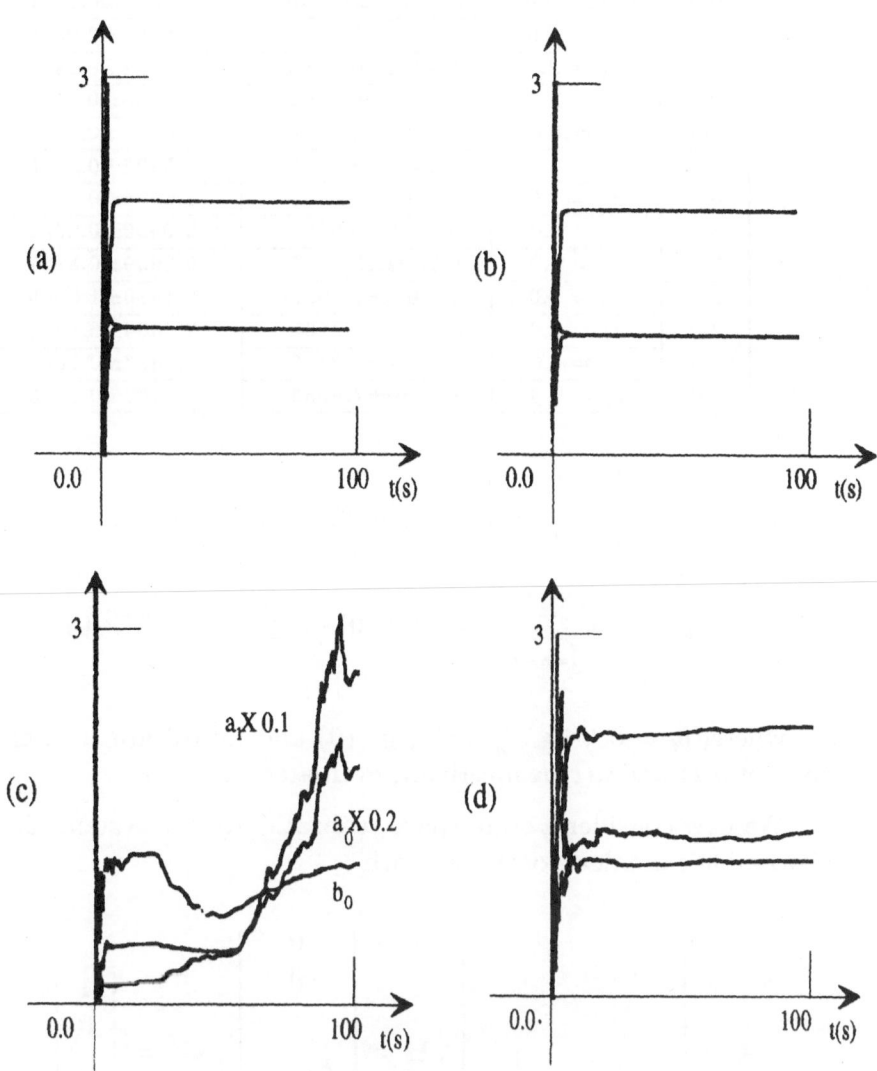

Figure 3.4: Pattern of parameter convergence in Example 3.2. (a) LS, no noise; (b) GLS, no noise; (c) LS, SNR=3.3 and (d) GLS, SNR=3.3

Table 3.2: Parameter estimates in LS and GLS schemes in Example 3.2

SNR	True values	LS est.±std. dev.	GLS est.±std. dev.
10.0	$a_0 = 1.0$	2.7116±0.3603	1.0202±0.0582
10.0	$a_1 = 2.0$	2.4764±0.1286	1.9994±0.0460
10.0	$b_0 = 1.0$	0.9942±0.0192	0.9958±0.0142
10.0	$b_1 = 0.0$	0.7025±0.1535	0.0081±0.0206
3.3	$a_0 = 1.0$	16.5390±5.3268	1.5402±0.9690
3.3	$a_1 = 2.0$	6.2767±1.6556	2.0585±0.2410
3.3	$b_0 = 1.0$	0.9545±0.0257	0.9600±0.0631
3.3	$b_1 = 0.0$	6.3692±2.2555	0.2059±0.3727
2.0	$a_0 = 1.0$	41.092±18.685	2.1510±1.5065
2.0	$a_1 = 2.0$	12.717±5.3763	2.0504±0.1832
2.0	$b_0 = 1.0$	0.8914±0.0466	0.8883±0.1652
2.0	$b_1 = 0.0$	16.536±7.8083	0.4187±0.5295

The initial condition terms γ_l in (3.2) are given by

$$\gamma_l = y^{(l)}(0) + \sum_{i=n-l}^{n-1} a_i\, y^{(i-n+l)}(0) - \sum_{i=n-l}^{n-1} b_i\, u^{(i-n+l)}(0), \qquad (3.35)$$

where, $b_i = 0, \forall i \in \mathcal{I}_{n-1}^{n-n_b+1}$, $y^{(l)}(0)$ and $u^{(l)}(0)$ are the l-th derivatives of $y(t)$ and $u(t)$ respectively, evaluated at $t = 0$.

We now consider a state space realization of the system (3.1) in the observable canonical form, i.e., with

$$\mathbf{A} = \begin{bmatrix} -a_{n-1} & \\ -a_{n-2} & \\ \vdots & \mathbf{I} \\ -a_1 & \\ -a_0 & \mathbf{0} \end{bmatrix}, \mathbf{B} = \begin{bmatrix} 0 \\ 0 \\ \vdots \\ b_{n-n_b} \\ \vdots \\ b_0 \end{bmatrix}, \mathbf{C}^T = \begin{bmatrix} 1 \\ 0 \\ \vdots \\ 0 \end{bmatrix}. \qquad (3.36)$$

It may be easily verified that in this canonical form, the initial values of the state and the unknowns γ_i are related by

$$x_i(0) = \gamma_{i-1}, \forall i \in \mathcal{I}_n.$$

Further,

$$x_i^{(n-i+1)}(t) = -\sum_{l=0}^{n-i} a_l \, y^{(l)}(t) + \sum_{l=0}^{n-i} b_l \, u^{(l)}(t), \forall \, i \in \mathcal{I}_n \qquad (3.37)$$

where $x_i^{(j)}(t)$ is the j-th derivative of the i-th element of $\mathbf{x}(t)$.

Integrating (3.37) $(n-i+1)$-times,

$$
\begin{aligned}
x_i(t) &= -\sum_{l=0}^{n-i} a_l \, y_{(n+1-i-l)}(t) + \sum_{l=0}^{n-i} b_l \, u_{(n+1-i-l)}(t) \\
&\quad + \sum_{l=i-1}^{n-1} \gamma_i \, \iota_{(i-l+1)}(t)
\end{aligned}
\qquad (3.38)
$$

Equation (3.38) with the estimated parameters and the initial state can be used for state estimation. It may, however, so happen that the monotonically increasing terms $\iota_{(l)}(t)$ blow up to large values for large t. This can be avoided by periodically resetting the integrals and correspondingly shifting the time origin. Based on this, the following recursive least squares algorithm for simultaneous estimation of parameters and state is derived. In order to obtain complete recursification in time, the BPF characterization is employed.

For $k = 1, 2, \ldots, m$ do

$$
\left.
\begin{aligned}
\hat{e}(k+1) &= y(k+1) - \phi_{1,k+1}^T \, \hat{\mathbf{p}}(k) \\
\mathbf{q}(k+1) &= \frac{\mathbf{P}(k) \, \phi_{1,k+1}}{[\rho(k+1) + \phi_{1,k+1}^T \, \mathbf{P}(k) \, \phi_{1,k+1}]} \\
\hat{\mathbf{p}}(k) &= \hat{\mathbf{p}}(k) + \mathbf{q}(k+1) \, \hat{e}(k+1) \\
\mathbf{P}(k+1) &= [\mathbf{P}(k) - \mathbf{q}(k+1) \, \phi_{1,k+1}^T \, \mathbf{P}(k)] / \rho(k+1) \\
\hat{x}_i(k) &= -\sum_{l=0}^{n-i} \hat{a}_l \, y_{(n+1-i-l)}(k) + \sum_{l=0}^{n-i} \hat{b}_l \, u_{(n+1-i-l)}(k) \\
&\quad + \sum_{l=i-1}^{n-1} \hat{\gamma}_i \, \iota_{(i-l+1)}(k), \forall \, i \in \mathcal{I}_n
\end{aligned}
\right\} \quad (3.39)
$$

where,

$$\hat{\mathbf{p}}^T = [\, \hat{a}_0 \ \hat{a}_1 \ \cdots \ \hat{a}_{n-1} \mid \hat{b}_0 \ \hat{b}_1 \ \cdots \ \hat{b}_{n_b} \mid \hat{\gamma}_0 \ \hat{\gamma}_1 \ \cdots \ \hat{\gamma}_{n-1} \,]$$

and $\hat{x}_i(k)$ is the estimate of the k-th BPF coefficient of $x_i(t)$ and $y_{(l)}(k), u_{(l)}(k)$ and $\iota_{(l)}(k)$ are the k-th BPF coefficients of the signals $y_{(l)}(t), u_{(l)}(t)$ and $\iota_{(l)}(t)$ respectively. They are reset periodically to zero with a resetting interval m_r, along with the following:

$\mathbf{P}(k) = \sigma \mathbf{I}, \sigma$ being a large positive constant, and

$\hat{\gamma}_{i-1} = \hat{x}_i(k), \forall i \in \mathcal{I}_n.$

Example 3.3. Simultaneous state and parameter estimation

Let us consider the second order system of Example 3.2. The input-output data shown in Fig. 3.3(a) & (c) is now used for simultaneous state and parameter estimation. Figures 3.5(a)–(d) show the pattern of parameter convergence for four values of SNR. The resetting interval m_r is chosen to be 100 and $\sigma = 10^4$. The convergence of the two state estimates $\hat{x}_1(t)$ and $\hat{x}_2(t)$ to the true states is shown in Fig. 3.6(a) and (b) respectively for SNR=3.3. The GLS method is employed (with $D(s) = s^n$) since the LS method would give biased parameter estimates in presence of noise. Reasonably good convergence of parameters and state is obtained despite the high level of noise.

3.8   Remarks

The framework of GHOF has been applied to the parameter estimation and combined parameter and state estimation problems of continuous-time systems. Keeping in view the need of real-time applications, the algorithms have been made recursive. In this context it was found that complete time-recursive forms can be obtained only in the special case of block pulse functions, while in the general case, block-recursive formulations were derived. In practical situations, where the execution time of the algorithms becomes critical, one may retain as many terms as possible in the GHOF framework. This therefore gives an additional degree of flexibility to the designer of an on-line algorithm in the choice of the number of terms suiting the problem at hand. Keeping this in view, we have used the single term approximation in the problem of simultaneous state and parameter estimation.

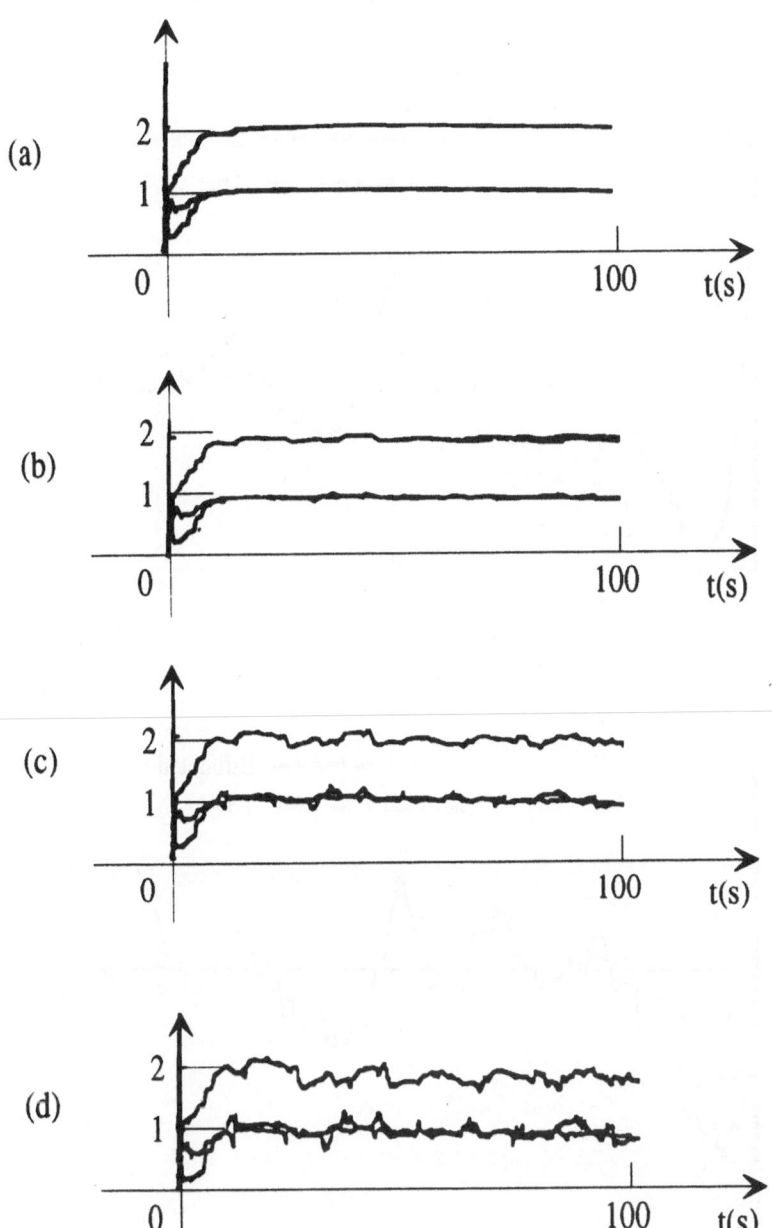

Figure 3.5: Parameter convergence in Example 3.3. (a) no noise; (b) SNR=10.0; (c) SNR=5.0 and (d) SNR=3.3

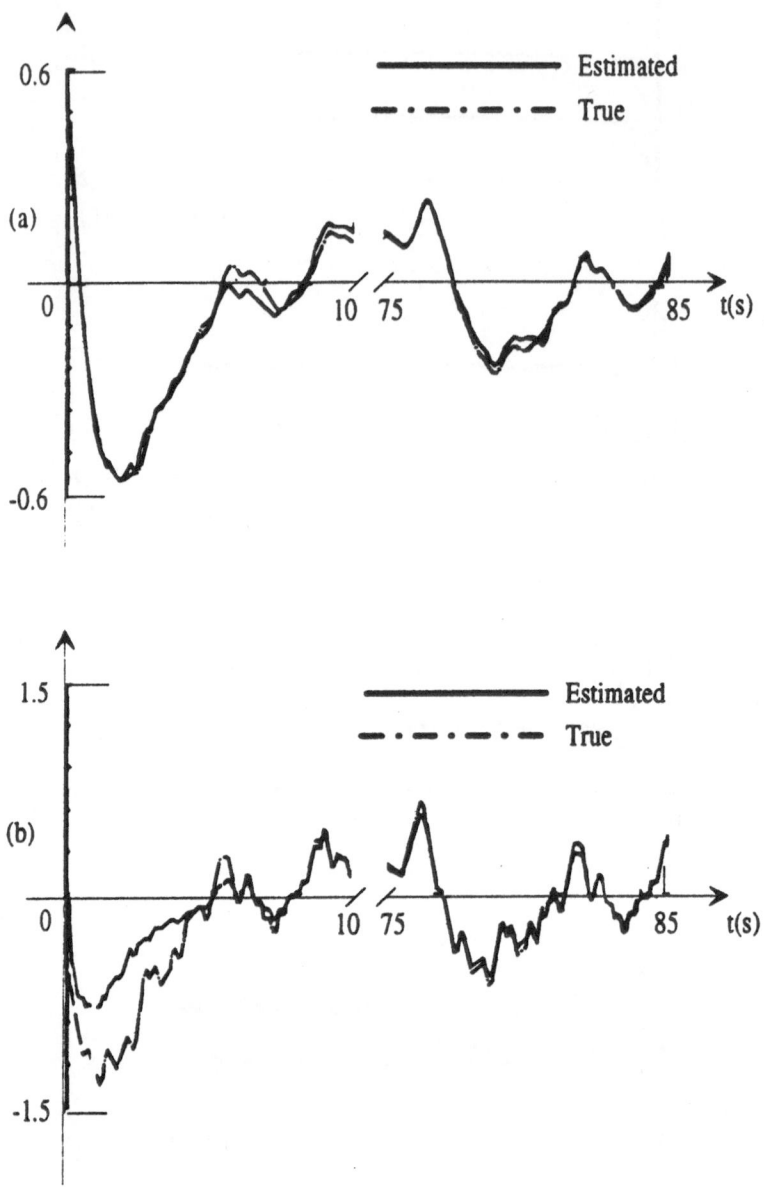

Figure 3.6: Convergence of state in Example 3.3 for SNR=3.3. (a) $x_1(t)$ and (b) $x_2(t)$

Chapter 4

Continuous-time Model-based Self-tuning Control

4.1 Survey of literature in the field

Self-tuning control (STC) has been an area of intense research activity during the past two decades. In 1965 Young reported on a completely continuous-time method of self-tuning control which involved implementation via analog hardware [306,307]. The first self-tuning control scheme based on a discrete-time (DT) model was proposed by Åström and Wittenmark in 1973 [6]. This led to an enormous amount of activity in the field based mostly on DT models [4]. In the subsequent period of time, which is known for great advances of digital computer technology, Young's CT-treatment went virtually unnoticed. Although the model reference adaptive control (MRAC) schemes were often formulated using CT models, these were rather simple in nature because of the need for realization via analog components.

STC and MRAC were considered to have separate identities until Egardt brought out the underlying equivalence between them in 1979 [70,71]. Egardt also formulated the self-tuning control problem in a CT framework keeping in view the vast developments in the DT approaches. This led to a revival of interest and activity in CT approaches. Gawthrop [77] suggested *hybrid* self-tuning control which considered continuous-time models for design and discrete-time techniques for implementation, and showed that it is superior to the completely digital scheme. He also unified a number of algorithms involving model reference, pole placement and predictive control [78]. Pointing out the problems associated with DT techniques mentioned in section 3.1, he advocated the use of CT design for STC. Some of the problems related to DT approaches have been recently solved by using the so-called *delta-operator* in the implementation of STC schemes [83].

Recently more complex self-tuning control schemes have been formulated. A number of schemes employing pole-zero placement have been proposed. The concept of an *emulator* is introduced by Gawthrop [79,81] and used to unify a number of existing algorithms and generate some new ones. Goodwin and Mayne [84] showed how the CT MRAC can be decomposed into separate stages concerned with parameter estimation and control giving rise to the *explicit* or *indirect* schemes in contrast to the *implicit* or *direct* schemes in vogue earlier. In the *implicit* schemes the second stage is avoided by a suitable reparametrization of the estimator so that the controller parameters are directly estimated. The explicit methods are more general since any parameter estimation scheme (e.g., Least Squares, Maximum Likelihood, gradient etc.) can be coupled with any controller design scheme (e.g., pole placement, optimal, LQG etc.) in the choicest combination.

The stability properties of the implicit CT STC schemes were studied by Egardt [70,72], Gawthrop [77,82], Elliott [73]. Explicit schemes have been considered more recently by Kreisselmeier [141], Goodwin and Middleton [85] and Middleton [170]. Similar studies in the MRAC context were reported by Narendra et al. [184] and Morse [176]. The robustness of CT schemes to unmodelled plant dynamics was questioned by Rohrs et al. [234] and some attempts to solve this problem are reported [81,82,171,3].

Table 4.1 summarizes the various CT approaches to STC indicating the salient features of the respective techniques.

In this chapter we consider an explicit self-tuning scheme with pole-placement design which uses characterization via orthogonal functions to retain the CT model parameters. To obtain a fully time-recursive formulation with minimum computational delay, the BPF characterization is used. Since it is also possible to implement controller action based on BPF, in real-time, a totally CT model-based STC scheme is presented.

Section 4.2 describes the STC problem with a general objective of

Table 4.1: Summary of continuous-time approaches to self-tuning control

Ref.	Year	Design	Impl.[1]	Scheme	Dist.[2]	Comments
[306, 307]	1965–66	CT	CT	Implicit	No	The first CT approach in self-tuning; Implementation via analog hardware
[70, 71, 72]	1979–80	CT	–	Implicit	No	Unification of MRAC and STC; Formulation of STC in CT; Analysis of stability
[77]	1980	CT	DT	Implicit	Yes	The *hybrid* approach
[184, 176]	1980	CT	–	Implicit	No	Stability analysis of CT MRAC
[73]	1982	CT	DT	Implicit	No	Hybrid approach to MRAC
[78]	1982	CT	–	Implicit	Yes	Unification of many adaptive control schemes; Justification for design using CT
[141]	1985	CT	–	Explicit	No	Analysis of stability
[234]	1985	CT	DT	Implicit	Yes	Questioned the robustness of CT MRAC schemes to unmodelled dynamics
[80]	1986	CT	DT	Implicit	Yes	Practical PID STC example in CT
[79]	1986	CT	–	Implicit	Yes	Unification of various designs using the concept of *emulator*
[83]	1986	DT	DT	Explicit	No	Use of *delta* operator to remove some difficulties of DT algorithms
[84]	1987	CT	–	Explicit	No	Global convergence with least squares parameter estimation
[170]	1987	CT	–	Explicit	No	Extension of [141]
[85]	1987	CT/DT	DT	Explicit	No	Unified CT/DT convergence analysis
[81]	1987	CT	DT	Implicit	Yes	Design of CT STC
[82]	1987	CT	DT	Implicit	Yes	Robustness of CT STC
[171]	1988	CT	–	Explicit	Yes	Design guidelines for STC
[203]	1989	CT	CT	Explicit	Yes	Completely CT model-based implementation

set-point tracking and disturbance rejection (both measurable and unmeasurable type) in the presence of parameter uncertainty. The implementation aspects of the STC scheme using the BPF approach are
described in section 4.3. Simulation experiments have been undertaken
to compare the performance of the proposed algorithm with that of a
DT scheme given in [62] and the results are summarized in Example 4.1.
Finally discussions on the results are made in section 4.4.

4.2   The STC problem in a CT setting

A continuous-time plant, with disturbances, can in general be modelled
by

$$A(s)Y(s) = B(s)U(s) + C(s)Z(s) + D(s)V(s), \qquad (4.1)$$

where $u(t)$ and $y(t)$ are the input and output of the plant and $z(t)$
and $v(t)$ are the unknown (unmeasurable) and known (measurable) disturbances respectively. For simplicity, the plant has been assumed to be
delay-free. A, B, C and D are polynomials of the form:

$$\left. \begin{aligned}
A(s) &= s^n + a_{n-1}s^{n-1} + \cdots + a_1 s + a_0 \\
B(s) &= b_{n_b}s^{n_b} + b_{n_b-1}s^{n_b-1} + \cdots + b_1 s + b_0 \\
C(s) &= s^n + c_{n-1}s^{n-1} + \cdots + c_1 s + c_0 \\
D(s) &= d_{n_d}s^{n_d} + d_{n_d-1}s^{n_d-1} + \cdots + d_1 s + d_0
\end{aligned} \right\} . \qquad (4.2)$$

The control objective is to make the system output $y(t)$ follow a
specified trajectory $w(t)$, while at the same time rejecting the effects
due to $z(t)$ and $v(t)$, in the presence of uncertainty in the coefficients of
A, B, C and D. It may be possible to attribute $z(t)$ to noise and $v(t)$ to
load disturbances in the plant.

The controller action is given by

$$F(s)Y(s) + G(s)U(s) + S(s)V(s) = H(s)W(s), \qquad (4.3)$$

corresponding to the closed loop system shown in Fig. 4.1.

The closed loop equations are

$$Y(s) = \frac{BH}{AG+BF}W(s) + \frac{CG}{AG+BF}Z(s) + \frac{DG-BS}{AG+BF}V(s) \quad (4.4)$$

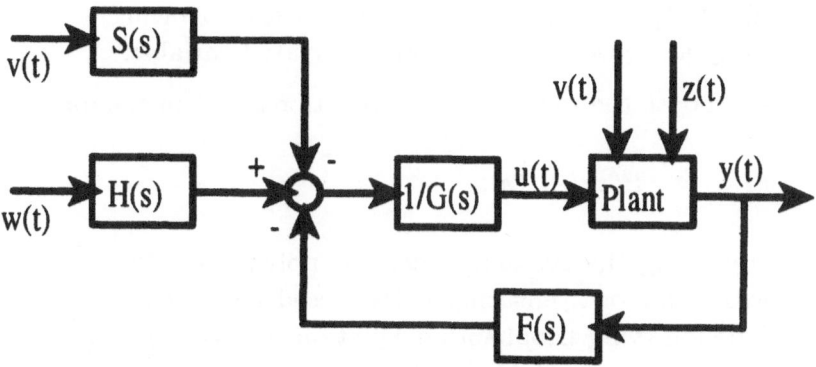

Figure 4.1: The structure of the continuous-time self-tuning controller

and

$$U(s) = \frac{AH}{AG + BF} W(s) - \frac{CF}{AG + BF} Z(s) - \frac{AS + DF}{AG + BF} V(s). \quad (4.5)$$

The description of the controller given here follows closely the scheme of [62] for a corresponding discrete-time controller.

To obtain a desired closed loop transfer function according to a reference model

$$Y(s) = M(s) W(s), \quad (4.6)$$

the design problem is to choose appropriate polynomials F, G, H and S satisfying (4.6). It may be noted here that this control scheme does not try to match the output $y(t)$ to the high frequency components of $w(t)$. This leads to better robustness properties of the scheme [81]. In the presence of $z(t)$ and $v(t)$, it may not be possible to achieve exact model following, but the design must be such that these disturbances have a minimal effect on the system output $y(t)$.

In a self-tuning framework, where the system polynomials are not exactly known or are time-varying in nature, it is necessary to estimate either $\{A, B, C, D\}$ (explicit approach) or $\{F, G, H, S\}$ (implicit approach) by measuring and processing $\{u(t), y(t), v(t)\}$ and estimating $z(t)$. We take up the explicit approach in which it is assumed that the estimates of $\{A, B, C, D\}$ are available as the output of a suitable parameter estimation algorithm and they are used in the computation of $\{F, G, H, S\}$ using pole placement. In the following, estimates are denoted by a ˆ above the symbol of the related variable.

If we choose the model transfer function $M(s)$ in the form

$$M(s) = \lambda B(s)/A_m(s), \qquad (4.7)$$

by retaining the system numerator polynomial $B(s)$ in the closed loop, with a factor λ ensuring unity closed loop steady state gain, we obtain the following Diophantine equation solvable for F and G [61]:

$$\hat{A}G + \hat{B}F = A_m \hat{C} \qquad (4.8)$$

and

$$H = \lambda \hat{C}, \qquad (4.9)$$

where $\lambda = A_m(0)/B(0)$.

The feedforward polynomial S is given by

$$\hat{D}G - \hat{B}S = \mu Q, \qquad (4.10)$$

where μ is arbitrarily small and Q is a polynomial in s such that $Q(0) = 0$ to ensure zero steady-state effect of the load disturbance $v(t)$ on the output. The value of μ may be chosen to be zero, but this may give rise to excessively high control action.

Equation (4.8) can be solved either by matrix inversion [62] or Euclid's algorithm [81] also applicable to discrete-time cases. The latter is a recursive algorithm very suitable for real-time applications.

Based on the above formulation, an algorithm with the following steps may be implemented:

Step 1: Initialize $\{\hat{A}, \hat{B}, \hat{C}, \hat{D}\}$ based on *a priori* knowledge;

Step 2: Calculate $\{F, G, H, S\}$ from equations (4.8–4.10);

Step 3: Calculate control signal $u(t)$ from equation (4.3) using a suitable numerical technique;

Step 4: Measure $y(t)$ and $v(t)$;

Step 5: Update the parameter estimates $\{\hat{A}, \hat{B}, \hat{C}, \hat{D}\}$ using a suitable parameter estimation algorithm;

Step 6: Go to Step 2.

4.3 Implementation of CT model-based STC

Steps 1, 2 and 4 of the algorithm presented in section 4.2 are common for both CT and DT techniques. It is in steps 3 and 5 that the two approaches differ. While in the conventional DT schemes step 3 would be performed by computing the values of $u(t)$ at different sampling instants corresponding to a suitable discretization of (4.5), in the CT scheme proposed here, (4.5) is characterized by orthogonal functions and the spectra of $u(t)$ are calculated based on those of $w(t)$, $\hat{z}(t)$ and $v(t)$. Since an adaptive controller has to operate in real-time, it is necessary to perform this computation in a time-recursive manner. Even though this is theoretically possible in the GHOF framework, due to restrictions on the time required for computation and collection of information to evaluate the spectra, the use of block pulse functions (BPF) for this purpose is preferred. Therefore, the SSST formula presented in section 2.4 is used to simulate the transfer functions in (4.5). To reduce the computational burden further, it may be simplified to

$$U(s) = \lambda \frac{A}{A_m} W(s) - \frac{F}{A_m} \hat{Z}(s) - \frac{AS + DF}{A_m \hat{C}} V(s). \tag{4.11}$$

Step 5 of the algorithm involves parameter estimation of a CT model which should also be performed recursively in time. This is achieved by employing the methodology presented in section 3.6. The GLS estimation algorithm is used because of the presence of the disturbance term $z(t)$. The recursive relation from which parameter estimation is to be performed is now of the form

$$z(t) = y(t) - \phi^T(t)\mathbf{p} \tag{4.12}$$

where,

$$\begin{aligned}\phi^T(t) = \ &[\, y_{(n)}(t) \mid \cdots \mid y_{(1)}(t) \mid u_{(n)}(t) \mid \cdots \mid u_{(n-n_b)}(t) \mid \\ &\hat{z}_{(n)}(t) \mid \cdots \mid \hat{z}_{(1)}(t) \mid v_{(n)}(t) \mid \cdots \mid v_{(n-n_d)}(t) \,]\end{aligned}$$

and

$$\mathbf{p}^T = [\, a_0 \;\cdots\; a_{n-1} \mid b_0 \;\cdots\; b_{n_b} \mid c_0 \;\cdots\; c_{n-1} \mid d_0 \;\cdots\; d_{n_d} \,].$$

In the above formulation the initial conditions are ignored with the assumption that the self-tuner starts when the plant is started with zero initial conditions.

The spectral approximations of the signals and their integrals in (4.12) which are obtained for successive time intervals are processed by the parameter estimation algorithm (3.20). To improve the robustness of the scheme several modifications have been suggested in the literature such as covariance resetting, projection of estimated parameters into a convex plane, coding the parameter estimation algorithm with square-root or UD factorization techniques, injection of perturbation signals for sufficient excitation of the plant etc. [61,86,171,305]. Other practical restrictions such as actuator saturation can also be taken into account in the simulation as will be seen in the following example.

Example 4.1. Performance of the CT self-tuning controller

Let us consider a system represented by

$$(s^2 + 2s + 1)Y(s) = U(s) + (0.1 + 0.1s + s^2)Z(s) + 0.2V(s).$$

It is to follow a unit amplitude square wave set-point signal of period 40s. The reference model is chosen as

$$(1 + 0.95s + 0.44s^2)Y(s) = W(s),$$

corresponding to a system with a bandwidth of 1.5 rad/s and a damping factor of 0.707. A load disturbance $v(t)$ in the form of a square wave of amplitude 0.5 and period 20s is assumed. The noise $z(t)$ is generated as a zero-mean Gaussian sequence.

To start with, the parameters are assumed to be only approximately known. We take $\hat{a}_0 = 0.5$, $\hat{a}_1 = 1.0$, $\hat{b}_0 = 0.5$, $\hat{c}_0 = 0.05$, $\hat{c}_1 = 0.05$ and $\hat{d}_0 = 0.1$. The covariance matrix is initialized with $\sigma = 10^4$.

The value of μ is taken to be zero. The Diophantine equation (4.8) is solved directly by matrix inversion.

To assess the performance of the proposed scheme, a standard DT algorithm [62] with the same controller structure is also implemented side by side. In this, the system is discretized by means of Tustin transformation to give rise to the following discrete equation:

$$\begin{aligned}
(1 - 1.80952q^{-1} + 0.81859q^{-2})Y(q) &= \\
0.00227(1 + 2q^{-1} + q^{-2})U(q) &+ \\
(0.91179 - 1.81361q^{-1} + 0.90272q^{-2})Z(q) &+ \\
0.000453(1 + 2q^{-1} + q^{-2})V(q)
\end{aligned}$$

where, q^{-1} is the backward shift operator.

Both the algorithms are subjected to the following three conditions:

case (i) $z(t) = 0$ and persistent excitation is maintained in the form of a PRBS signal of amplitude ± 0.05 units and period 127 riding over the control signal $u(t)$;

case (ii) $z(t) = 0$, but the PRBS perturbation signal is removed, and

case (iii) Non-zero $z(t)$, with perturbation signal added as in case (i). This situation corresponds to an SNR (as defined in Example 3.2) of approximately 11.0.

Furthermore, the control signal is clipped at ± 5 units to prevent it from being unbounded (in case the self-tuner fails) or to take into account actuator saturation. Before making the self-tuner on-line, the parameter estimation is allowed to proceed for 10s and during that period the controller parameters are kept at their nominal values computed at the beginning using the initial parameter estimates. In practical cases a much more sophisticated software structure is required which supervises the operation of the self-tuning controller [305].

Figures 4.2–4.5 show the results obtained using the DT and CT schemes. Both approaches work well in case (i). But it is observed that in cases (ii) and (iii) the DT scheme fails. In both cases the reason is the lack of parameter convergence. In case (ii), insufficient excitation causes oscillations in the control signal and even after this *oscillating excitation* finally leads to parameter convergence, the system has not been able to return to the desired operating point during the period of observation. Such a phenomenon was reported by Anderson [2]. But the CT scheme works quite well. In case (iii), the DT identification algorithm gives

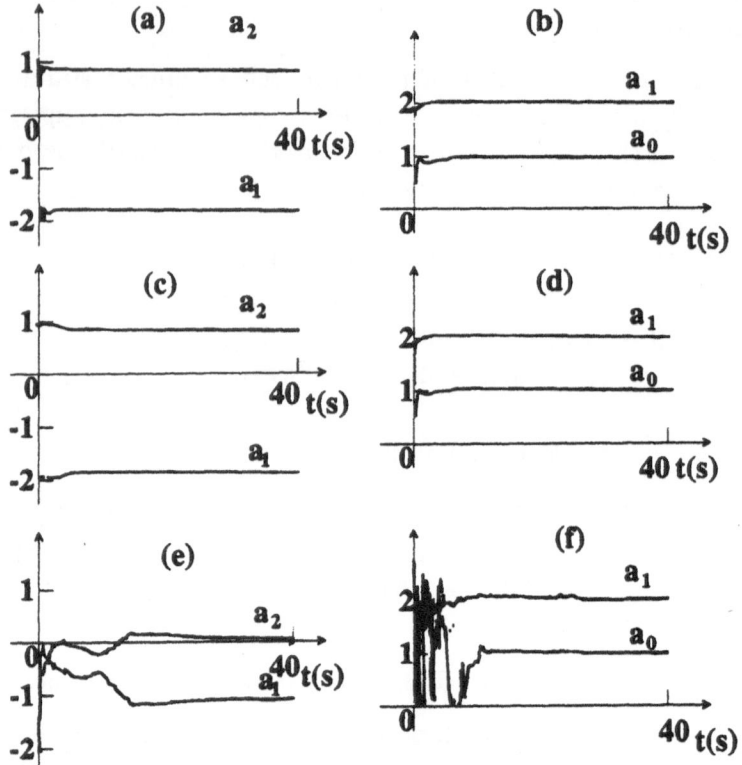

Figure 4.2: Estimates of a-parameters in Example 4.1 : (a) case (i) – DT, (b) case (i) – CT, (c) case (ii) – DT, (d) case (ii) – CT, (e) case (iii) – DT and (f) case (iii) – CT

biased parameter estimates due to noise and leads to clipping of the control signal at -5 units. However, the proposed algorithm works quite well in this case as well.

It has been observed that the coefficients of C do not necessarily converge to their true values. This is also the case in the DT context [62]. This does not, however, affect the control performance since $C(s)$ gets cancelled from the closed loop transfer function.

4.4   Remarks

The process of discretization of the originally CT plant often increases the number of unknown parameters, as seen in Example 4.1. Therefore,

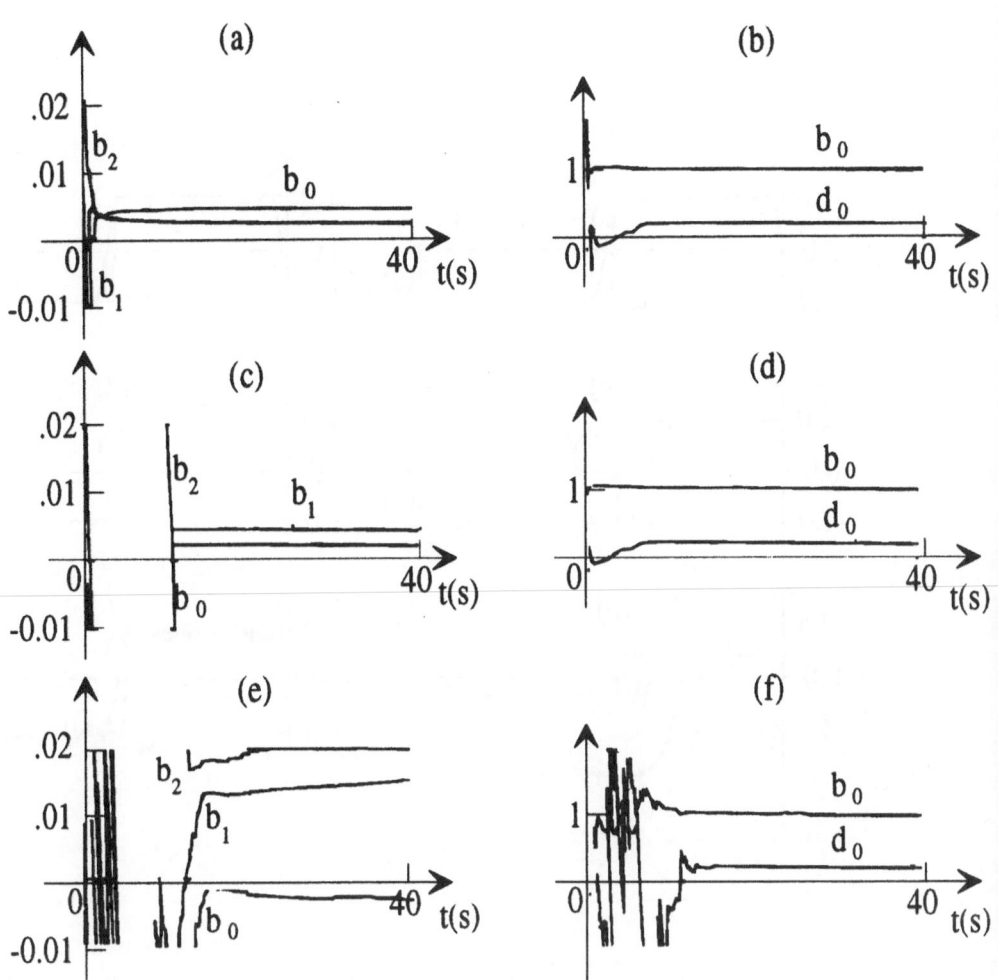

Figure 4.3: Estimates of b-parameters in Example 4.1 : (a) case (i) – DT, (b) case (i) – CT, (c) case (ii) – DT, (d) case (ii) – CT, (e) case (iii) – DT and (f) case (iii) – CT

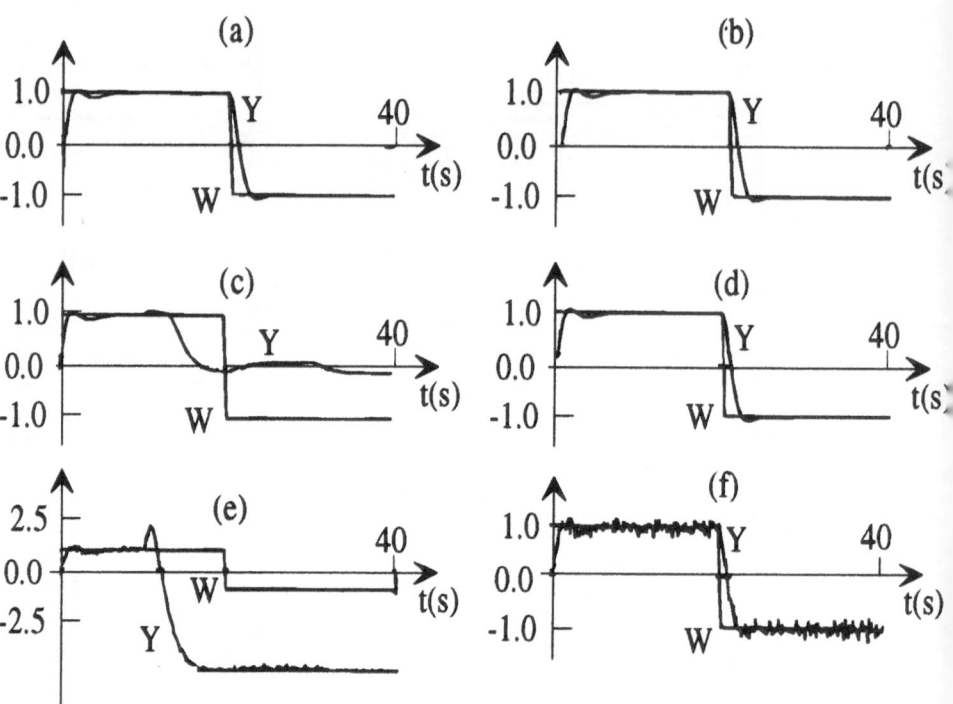

Figure 4.4: Output signal $y(t)$ and set-point $w(t)$ in Example 4.1 : (a) case (i) – DT, (b) case (i) – CT, (c) case (ii) – DT, (d) case (ii) – CT, (e) case (iii) – DT and (f) case (iii) – CT

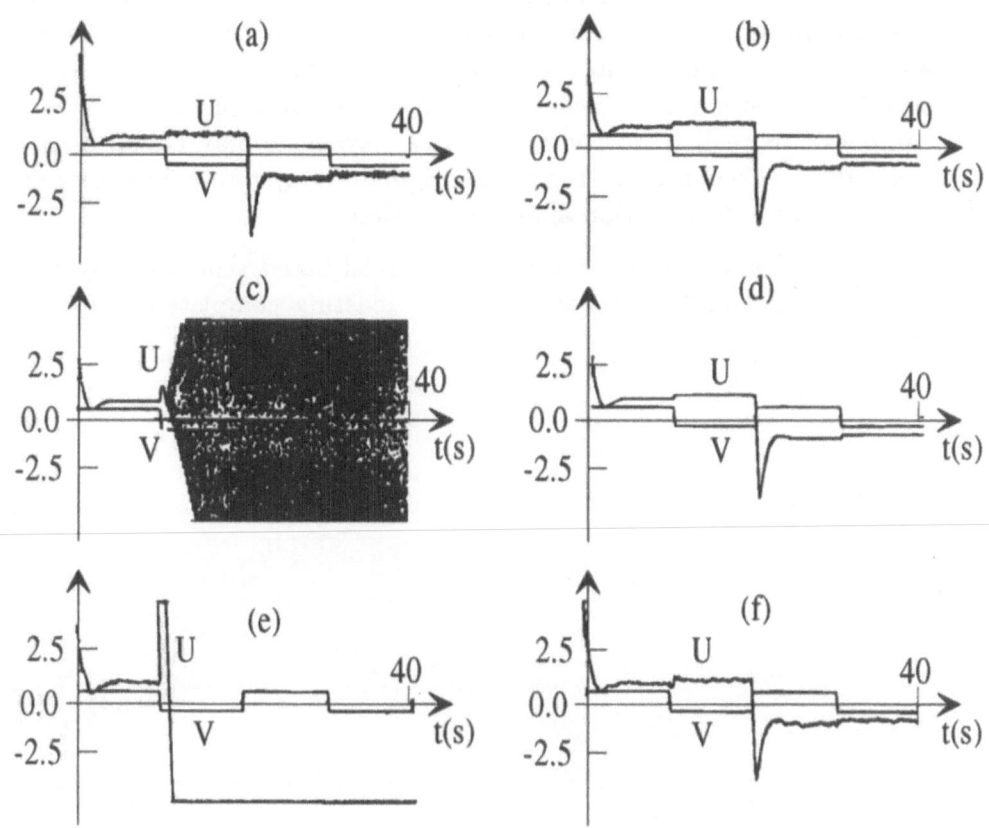

Figure 4.5: Control signal $u(t)$ and disturbance $w(t)$ in Example 4.1 : (a) case (i) – DT, (b) case (i) – CT, (c) case (ii) – DT, (d) case (ii) – CT, (e) case (iii) – DT and (f) case (iii) – CT

the parameter estimation algorithm needs inputs with a larger number of frequency components than is required by the CT scheme. In presence of noise, the CT parameter estimation algorithm, which inherently uses integration of input-output data, is seen to have superior performance. The choice of sampling interval is also not a trivial problem. Discretization, which transforms the semi-infinite left-half of the s-plane into a finite unit disc in the z-plane, often leads to clustering of poles and zeros, especially at higher sampling rates. Although, in theory, the transformation is fully justified for convenience of implementation in a computer, in reality, however, numerical problems may come up. At least, these effects are averted if the original CT system parameters are retained in the formulation and computation.

Because of the above reasons CT model-based schemes may now be viewed as viable alternatives to the existing completely DT STC schemes.

Chapter 5

Other Possible Applications

The framework of GHOF proposed here encompasses all the known systems of orthogonal functions, rendering the resulting basis quite general, flexible and thereby suitable to approximate a wide class of square integrable functions in a variety of situations in the actual practice. The set of GHOF is shown to be superior to the existing ones in approximating discontinuous signals and in the analysis of dynamical systems subjected to such inputs.

The selected applications of the proposed set of GHOF illustrated here are prompted by the special nature of certain problems encountered frequently in practice. In these cases it is shown how the definition of the set of GHOF can be tailored according to the needs of the problem under consideration. The effectiveness of the set of GHOF as an appropriate basis in the analysis and identification of power electronic circuits and systems and in the study of highly nonlinear Van der Pol oscillators along with its reducibility to any of the existing systems of orthogonal functions amply demonstrated in Chapters 2 and 3 clearly suggests that the proposed system has potential for further applications. The case of continuous-time model based self-tuning control is handled with particular reference to block-pulse functions only with a view to show the possibility for real-time implementation at the same time maintaining the relationship with the GHOF.

In contrast with the coverage of applications with those in the book by Rao [216], this book does not include the following:

(a) Analysis of time-delay systems

(b) Solution of functional differential equations

(c) Optimal control of lumped linear systems with and without delay terms

(d) Distributed parameter systems.

Having set out in detail with the derivation of the basic operational matrix for integration via GHOF and its use in the solution of state equation, some marginal amount of exercise on part of the reader will pave way towards the solution of the above problems as outlined below:

(i) Analysis of time-delay systems:

The essential requirement in this case is the formal derivation of the operational matrix for delay via GHOF. This can be accomplished in a straight forward manner by considering a delayed function and expanding it in terms of GHOF in a procedure well illustrated in [216]. This matrix along with the one already derived for integration converts the given state equation containing delay terms into an algebraic form which can be solved for the GHOF spectrum of the state or output as required.

(ii) Solution of functional differential equation with terms having stretched argument:

An operational matrix for stretch via GHOF can be derived as shown in [216]. This matrix and the one for integration become the means by which the system equation is converted into algebraic form for solution.

(iii) Optimal Control:

The solution of optimal control problems is obtained through the solution of the related state and costate equations. Depending on the nature of the system, use of the operational matrices for integration and delay provides solution to a wide class of problems. That is, the tools developed in this book together with those proposed in (i) above are sufficient to give solutions to problems of optimal control for systems with or without delays.

(iv) Distributed parameter systems:

As has been illustrated in [216], to handle partial differential equations, operational matrices for integration with respect to the various independent variables of space and time can be derived by first defining GHOF in more than one dimension. This exercise too is straightforward.

The resultant operational matrices would reduce the partial calculus of distributed parameter systems to an algebra. The solutions of partial differential equations can be obtained in terms of the multidimensional GHOF spectra.

Several other issues related to sensitivity, suboptimal control, interpretation of system properties such as stability, controllability, observability etc. can also be handled on the basis of the GHOF spectral description of systems. The central idea behind all these and any other problems not mentioned here, is the use of appropriate basis and the related operational matrices. This makes it clear as to where else the GHOF framework can be extended for useful application either for numerical computation or simulation in the study of systems and control.

However, the authors would like to caution the reader against over-enthusiastic and indiscriminate use of the GHOF framework in its general setting unless the situation calls for their imminent application. That is, the functions encountered or expected in the study should posses combined features of continuity and jumps or discontinuities for a full scale effort in the use of GHOF in their general setting.

The definition itself of the GHOF should be tailored to match the needs of the problem. One suggestion is to incorporate a general package of GHOF with the facility to *define* based on options chosen to suit the problem at hand. That is, for instance, if one wishes to choose any of the existing conventional sets of orthogonal functions over finite interval, a *single segment* version should be chosen.

A case worthy of further investigation is that of minimum time control in which piecewise constant control inputs feature in the solution. The case of sliding mode control is another. Thus, this book together with the one by Rao [216] should provide the basis for effectively solving a wide class of problems in Systems and Control in terms of orthogonal functions.

Bibliography

[1] M. Agarwal and C. Canudas. On-line estimation of time-delay and continuous-time process parameters. *Int. J. Cont.*, 46(1):295–311, 1987.

[2] B. D. O. Anderson. Adaptive systems, lack of persistency of excitation and bursting phenomena. *Automatica*, 21(3):247–258, 1985.

[3] B. D. O. Anderson et al. *Stability of Adaptive Systems, Passivity and Averaging Analysis*. MIT Press, 1986.

[4] K. J. Åström. Theory and applications of adaptive control – a survey. *Automatica*, 19(5):471–486, 1983.

[5] K. J. Åström, P. Hagander, and J. Sternby. Zeros of sampled systems. *Automatica*, 20(1):31–38, 1980.

[6] K. J. Åström and B. Wittenmark. On self-tuning regulators. *Automatica*, 9(2):185–199, 1973.

[7] D. P. Atherton. *Nonlinear Control Engineering*. Van Nostrand Reinhold, London, 1975.

[8] E. V. Bohn. Estimation of continuous-time linear system parameters from periodic data. *Automatica*, 18(2):27–36, Feb. 1982.

[9] E. V. Bohn. Measurement of continuous-time linear system parameters via Walsh functions. *IEEE Trans. IE*, IE–29(1):38–46, 1982.

[10] E. V. Bohn. Optimal Walsh function input signals for parameter identification in identification systems. *Math. & Comp. Simul.*, 27(5/6):485–490, Oct. 1985.

[11] E. V. Bohn. Walsh functions decoupled parameter estimation equations for dynamic continuous-time models with time-delay. In *Proc. 7th IFAC Symp. on Ident. & System Param. Estim.*, pages 799–802, University of York, U. K., July 1985. Pergamon Press.

[12] R. G. Cameron, B. Kouvaritakis, and S. Mossaheb. A new approach to the prediction of limit cycles. *Int. J. Cont.*, 32(6):963–981, 1980.

[13] R. G. Cameron and M. Tabatabai. Predicting the existence of limit cycles using Walsh functions : Some further results. *Int. J. Syst. Sci.*, 14(9):1043–1064, Sept. 1983.

[14] S. L. Campbell. On using orthogonal functions with singular systems. *IEE Proc., Part – D, CTA*, 131(6):267–268, Nov. 1984.

[15] B. Chakravarty, N. Mandayam, S. Mukhopadhyay, A. Patra, and G. P. Rao. Real-time parameter estimation via block pulse functions. In *SICE '89*, pages 1095–1098, Matsuyama, Japan, July 1989.

[16] R. Y. Chang, K. C. Chen, and M. L. Wang. A new approach to the parameter estimation of linear time-varying delayed systems via modified Laguerre polynomials. *Int. J. Syst. Sci.*, 16(12):1505–1515, Dec. 1985.

[17] R. Y. Chang, C. K. Chou, and M. L. Wang. Solution of functional differential equations via generalized block pulse functions. *Int. J. Syst. Sci.*, 16(11):1431–1440, Oct. 1985.

[18] R. Y. Chang and M. L. Wang. Parameter identification via shifted Legendre polynomials. *Int. J. Syst. Sci.*, 13(10):1125–1135, Oct. 1982.

[19] R. Y. Chang and M. L. Wang. Model reduction and control system design by shifted Legendre polynomial functions. *J. Dynamic Syst. Measur. & Cont.*, 105(1):52–55, Mar. 1983.

[20] R. Y. Chang and M. L. Wang. Analysis of stiff systems via method of shifted Legendre functions. *Int. J. Syst. Sci.*, 15(6):627–637, June 1984.

[21] R. Y. Chang and M. L. Wang. Legendre polynomials approximation to dynamic linear state equations with initial or boundary value conditions. *Int. J. Cont.*, 40(1):215–232, July 1984.

[22] R. Y. Chang and M. L. Wang. Solution of population balance equation of breakage model via shifted Legendre functions. *Int. J. Syst. Sci.*, 15(1):63–74, Jan. 1984.

[23] R. Y. Chang and M. L. Wang. The application of shifted Legendre polynomials to time-delay systems and parameter identification. *J. Dynamic Syst. Measur. & Cont.*, 107(1):79–85, Mar. 1985.

[24] R. Y. Chang and M. L. Wang. Solutions of integral equations via shifted Legendre polynomials. *Int. J. Syst. Sci.*, 16(2):197–208, Feb. 1985.

[25] R. Y. Chang and S. Y. Yang. Solution of two-point boundary value problems by generalized orthogonal polynomials and application to optimal control of lumped and distributed parameter systems. *Int. J. Cont.*, 43(6):1785–1802, June 1986.

[26] R. Y. Chang, S. Y. Yang, and M. L. Wang. A new approach for parameter identification of time-varying systems via generalized orthogonal polynomials. *Int. J. Cont.*, 44(6):1747–1755, Dec. 1986.

[27] R. Y. Chang, S. Y. Yang, and M. L. Wang. Solutions of linear dynamic systems by generalized orthogonal polynomials. *Int. J. Syst. Sci.*, 17(12):1727–1740, Dec. 1986.

[28] R. Y. Chang, S. Y. Yang, and M. L. Wang. Analysis of stiff systems via the method of generalized orthogonal polynomials. *Int. J. Syst. Sci.*, 18(1):97–116, Jan. 1987.

[29] R. Y. Chang, S. Y. Yang, and M. L. Wang. Solution of a scaled system via generalized orthogonal polynomials. *Int. J. Syst. Sci.*, 18(12):2369–2382, Dec. 1987.

[30] R. Y. Chang, S. Y. Yang, and M. L. Wang. Solution of integral equations via generalized orthogonal polynomials. *Int. J. Syst. Sci.*, 18(3):553–568, Mar. 1987.

[31] Y. C. Chao, C. L. Chen, and H. P. Huang. Recursive parameter estimation of transfer function matrix models via Simpson's integrating rules. *Int. J. Syst. Sci.*, 18(5):901–911, May 1987.

[32] C. F. Chen and C. H. Hsiao. A state-space approach to Walsh series solution of linear systems. *Int. J. Syst. Sci.*, 6(9):833–858, Sept. 1975.

[33] C. K. Chen and C. Y. Yang. Analysis and parameter identification of time-delay systems via polynomial series. *Int. J. Cont.*, 46(1):111–127, July 1987.

[34] W. L. Chen. Block pulse series analysis of scaled systems. *Int. J. Syst. Sci.*, 12(7):885–891, July 1981.

[35] W. L. Chen. Walsh series analysis of multi-delay systems. *J. Franklin Inst.*, 313(4):207–217, Apr. 1982.

[36] W. L. Chen and C. Y. Chung. Error analysis of block pulse series solutions. *Int. J. Syst. Sci.*, 17(12):1669–1676, Dec. 1986.

[37] W. L. Chen and C. Y. Chung. New integral operational matrix in block pulse series analysis. *Int. J. Syst. Sci.*, 18(3):403–408, Mar. 1987.

[38] W. L. Chen and C. S. Hsu. Convergence of the block pulse series solution of a linear time-invariant system. *Int. J. Syst. Sci.*, 15(4):351–360, Apr. 1984.

[39] W. L. Chen and B. S. Jeng. Analysis of piecewise constant delay systems via block pulse functions. *Int. J. Syst. Sci.*, 12(5):625–633, May 1981.

[40] W. L. Chen and C. L. Lee. On the convergence of the block pulse series solution of a linear time-invariant system. *Int. J. Syst. Sci.*, 13(5):491–498, May 1982.

[41] W. L. Chen and C. L. Lee. Walsh series expansion of composite functions and its application to linear systems. *Int. J. Syst. Sci.*, 13(2):219–226, Feb. 1982.

[42] W. L. Chen and J. F. Lin. Analysis and identification of systems with a nonlinear element. *Int. J. Syst. Sci.*, 17(7):1097–1104, July 1986.

[43] W. L. Chen and Y. P. Shih. Analysis and optimal control of time-varying linear systems via Walsh functions. *Int. J. Cont.*, 27(6):917–932, June 1978.

[44] W. L. Chen and Y. P. Shih. Parameter estimation of bilinear systems via Walsh functions. *J. Franklin Inst.*, 305(5):249–257, May 1978.

[45] W. L. Chen and Y. P. Shih. Shift Walsh matrix and delay differential equations. *IEEE Trans. AC*, AC–23(6):1023–1028, Dec. 1978.

[46] W. L. Chen and S. G. Wu. Analysis of multirate sampled-data systems by block pulse functions. *Math. & Comp. Simul.*, 27(5/6):503–510, Oct. 1985.

[47] W. L. Chen and S. G. Wu. Analysis of sampled-data systems by block pulse functions. *Int. J. Syst. Sci.*, 16(6):745–752, June 1985.

[48] W. L. Chen and S. G. Wu. Analysis and optimal control of PWM systems. *Int. J. Cont.*, 45(5):1565–1574, May 1987.

[49] B. Cheng and N. S. Hsu. Analysis and parameter estimation of bilinear systems via block pulse functions. *Int. J. Cont.*, 36(1):53–65, July 1982.

[50] B. Cheng and N. S. Hsu. Single input single output system identification via block pulse functions. *Int. J. Syst. Sci.*, 13(6):697–702, June 1982.

[51] J. H. Chou. Analysis and identification of scaled systems via shifted Jacobi series. *Int. J. Syst. Sci.*, 18(1):33–41, Jan. 1987.

[52] J. H. Chou and I. R. Horng. Chebyshev series analysis and identification of scaled systems. *Int. J. Syst. Sci.*, 16(9):1157–1162, Sept. 1985.

[53] J. H. Chou and I. R. Horng. Double-shifted Chebyshev series for convolution integral and integral equations. *Int. J. Cont.*, 42(1):225–232, July 1985.

[54] J. H. Chou and I. R. Horng. Identification of time-varying bilinear systems using Legendre series. *J. Franklin Inst.*, 322(5/6):353–359, 1986.

[55] J. H. Chou and I. R. Horng. Parameter identification of lumped time-varying systems via shifted Chebyshev series. *Int. J. Syst. Sci.*, 17(3):459–464, Mar. 1986.

[56] J. H. Chou and I. R. Horng. Shifted Chebyshev series analysis and identification of time-varying bilinear systems. *Int. J. Cont.*, 43(1):129–137, Jan. 1986.

[57] J. H. Chou and I. R. Horng. State estimation using continuous orthogonal functions. *Int. J. Syst. Sci.*, 17(9):1261–1267, Sept. 1986.

[58] J. H. Chou and I. R. Horng. Parameter identification of nonlinear systems via shifted Chebyshev series. *Int. J. Syst. Sci.*, 18(5):895–900, May 1987.

[59] H. Y. Chung. System identification via Fourier series. *Int. J. Syst. Sci.*, 18(6):1191–1194, June 1987.

[60] H. Y. Chung and Y. Y. Sun. Analysis and parameter estimation of scaled systems using the Taylor series approach. *CTAT*, 3(4):381–385, Dec. 1987.

[61] D. W. Clarke. Implementation of self-tuning controllers. In C. J. Harris and S. A. Billings, editors, *Self-tuning and Adaptive Control*, chapter 5. Peter Peregrinus, 1982.

[62] D. W. Clarke. Model following and pole placement self-tuners. *Opt. Cont. Appln. & Methods*, 3(4):323–335, 1982.

[63] P. R. Clement. Laguerre functions in signal analysis and parameter identification. *J. Franklin Inst.*, 313(2):85–95, Feb. 1982.

[64] M. S. Corrington. Solution of differential and integral equations with Walsh functions. *IEEE Trans. CT*, CT–20(5):470–476, Sept. 1973.

[65] N. K. De and A. K. Chattopadhyay. Modelling a dual-converter/DC motor drive. *Elect. Mach. & Electromechanics*, 5:407–416, 1980.

[66] A. Deb and A. K. Dutta. Analysis of pulse-fed power electronic circuits using Walsh functions. *Int. J. Electronics*, 62(3):449–459, Mar. 1987.

[67] B. Dwolatzky. A general solution to the system identification problem using intermediate domain processing. *Int. J. Cont.*, 36(3):493–510, Sept. 1982.

[68] B. Dwolatzky. Intermediate domain system identification using Walsh transform. *Automatica*, 20(2):237–242, Feb. 1984.

[69] J. M. Edmunds. Identifying sampled-data systems using difference operator models. Technical Report 601, UMIST, Manchester, U.K., 1984.

[70] B. Egardt. *Stability of Adaptive Controllers*, volume 20 of *LNCIS*. Springer Verlag, 1979.

[71] B. Egardt. Unification of some continuous-time adaptive control schemes. *IEEE Trans. AC*, AC–24(4):588–592, Aug. 1979.

[72] B. Egardt. Stability analysis of continuous-time adaptive control systems. *SIAM J. Cont. & Optim.*, 18(5):540–558, Sept. 1980.

[73] H. Elliott. Hybrid adaptive control of continuous-time systems. *IEEE Trans. AC*, AC–27(2):419–426, Apr. 1982.

[74] M. H. ElShafey and E. V. Bohn. Use of modal functions for continuous-time system identification and state observer design. *Int. J. Cont.*, 45(5):1723–1736, May 1987.

[75] W. Fei. Walsh functions approach to identification of time-lag systems. *J. North China Univ. Tech.*, 4(1), 1988. (In Chinese).

[76] F. Fnaiech and L. Ljung. Recursive identification of bilinear systems. *Int. J. Cont.*, 45(2):453–470, Feb. 1987.

[77] P. J. Gawthrop. Hybrid self-tuning control. *IEE Proc., Part – D, CTA*, 127(5):229–236, Sept. 1980.

[78] P. J. Gawthrop. A continuous-time approach to discrete-time self-tuning control. *Opt. Cont. Appln. & Methods*, 3(4):399–414, 1982.

[79] P. J. Gawthrop. Continuous-time self-tuning control – a unified approach. In *Proc. of IFAC Symp. on Adaptive Systems in Control and Signal Processing*, Lund Institute of Technology, Sweden, 1986. Pergamon Press.

[80] P. J. Gawthrop. Self-tuning PID controllers, algorithms and implementation. *IEEE Trans. AC*, AC–31(3):201–209, Mar. 1986.

[81] P. J. Gawthrop. *Continuous-time Self-tuning Control, Vol. 1 – Design*. Engg. Control Series. Research Studies Press, Lechworth, England, 1987.

[82] P. J. Gawthrop. Robust stability of a continuous-time self-tuning controller. *Int. J. Adapt. Cont. & Sig. Proc.*, 1(1):31–48, 1987.

[83] G. C. Goodwin, R. Lozano Leal, D. Q. Mayne, and R. H. Middleton. A rapprochement between continuous and discrete model reference adaptive control. *Automatica*, 22(2):199–207, 1986.

[84] G. C. Goodwin and D. Q. Mayne. A parameter estimation perspective of continuous-time model reference adaptive control. *Automatica*, 23(1):57–70, 1987.

[85] G. C. Goodwin and R. H. Middleton. Continuous and discrete adaptive control. In C. T. Leondes, editor, *Control and Dynamic Systems*, volume 25. Academic Press, 1987.

[86] G. C. Goodwin and K. S. Sin. *Adaptive Filtering, Prediction and Control*. Prentice Hall, 1984.

[87] U. W. Hochstrasser. Orthogonal polynomials. In M. Abramowitz and I. A. Stegun, editors, *Handbook of Mathematical Functions*. Dover Publications, New York, 1967.

[88] I. R. Horng and J. H. Chou. Analysis, parameter estimation and optimal control of time-delay systems via Chebyshev series. *Int. J. Cont.*, 41(5):1221–1234, May 1985.

[89] I. R. Horng and J. H. Chou. The design of optimal observers via shifted Chebyshev polynomials. *Int. J. Cont.*, 41(2):549–556, Feb. 1985.

[90] I. R. Horng and J. H. Chou. Analysis and parameter identification of time-delay systems via shifted Jacobi polynomials. *Int. J. Cont.*, 44(4):935–942, Oct. 1986.

[91] I. R. Horng and J. H. Chou. Design of optimal observers with specified eigenvalues via shifted Legendre polynomials. *J. Opt. Theory & Appln.*, 51(1):179–188, Oct. 1986.

[92] I. R. Horng and J. H. Chou. Analysis and identification of nonlinear systems via shifted Jacobi series. *Int. J. Cont.*, 45(1):279–290, Jan. 1987.

[93] I. R. Horng and J. H. Chou. Legendre series for the identification of nonlinear lumped systems. *Int. J. Syst. Sci.*, 18(6):1139–1144, June 1987.

[94] I. R. Horng, J. H. Chou, and T. W. Yang. Model reduction of digital systems using discrete Walsh series. *IEEE Trans. AC*, AC–31(10):962–964, Oct. 1986.

[95] I. R. Horng and S. J. Ho. Discrete Walsh operational matrices for analysis and optimal control of linear digital systems. *Int. J. Cont.*, 42(6):1443–1455, Dec. 1985.

[96] I. R. Horng and S. J. Ho. Discrete pulse orthogonal functions for the analysis, parameter estimation and optimal control of linear time-varying digital systems. *Int. J. Cont.*, 45(6):1975–1984, June 1987.

[97] I. R. Horng and S. J. Ho. Discrete pulse orthogonal functions for the analysis, parameter estimation and optimal control of linear digital systems. *Int. J. Cont.*, 45(2):597–605, Feb. 1987.

[98] N. S. Hsu and B. Cheng. Analysis and optimal control of time-varying linear systems via block pulse functions. *Int. J. Cont.*, 33(6):1107–1122, June 1981.

[99] C. Hwang. Solution of a functional differential equation via delayed unit step functions. *Int. J. Syst. Sci.*, 14(9):1065–1073, Sept. 1983.

[100] C. Hwang. Solution of a scaled system via Chebyshev polynomials. *J. Franklin Inst.*, 318(4):233–241, Oct. 1984.

[101] C. Hwang and C. M. Chen. Solution of a linear differential equation of the stretched type via Laguerre functions. *J. Franklin Inst.*, 323(3):385–394, Mar. 1987.

[102] C. Hwang, C. T. Chen, and Y. P. Shih. Parameter estimation of discrete systems via Hahn polynomials. *Int. J. Cont.*, 46(5):1605–1619, Nov. 1987.

[103] C. Hwang and M. Y. Chen. Analysis and optimal control of time-varying linear systems via shifted Legendre polynomials. *Int. J. Cont.*, 41(5):1317–1330, May 1985.

[104] C. Hwang and M. Y. Chen. Analysis and parameter identification of time-delay systems via shifted Legendre polynomials. *Int. J. Cont.*, 41(2):403–415, Feb. 1985.

[105] C. Hwang and M. Y. Chen. Parameter identification of bilinear systems using the Galerkin method. *Int. J. Syst. Sci.*, 16(5):641–648, May 1985.

[106] C. Hwang and M. Y. Chen. Analysis and parameter identification of bilinear systems via shifted Legendre polynomials. *Int. J. Cont.*, 44(2):351–362, Aug. 1986.

[107] C. Hwang and M. Y. Chen. Solution of a scaled system by shifted Legendre series representation. *Comp. & Elect. Engg.*, 12(1/2):13–22, Jan./Feb. 1986.

[108] C. Hwang and T. Y. Guo. Identification of lumped linear time-varying systems via block pulse functions. *Int. J. Cont.*, 40(3):571–583, Sept. 1984.

[109] C. Hwang and T. Y. Guo. New approach to the solution of integral equations via block pulse functions. *Int. J. Syst. Sci.*, 15(4):361–373, Apr. 1984.

[110] C. Hwang and T. Y. Guo. Parameter identification of a class of time-varying systems via orthogonal shifted Legendre polynomials. *J. Franklin Inst.*, 318(1):59–69, July 1984.

[111] C. Hwang and T. Y. Guo. Transfer function matrix identification in MIMO systems via shifted Legendre polynomials. *Int. J. Cont.*, 39(4):807–814, Apr. 1984.

[112] C. Hwang and Y. P. Shih. Laguerre series solution of a functional differential equation. *Int. J. Syst. Sci.*, 13(7):783–788, July 1982.

[113] C. Hwang and Y. P. Shih. Parameter identification via Laguerre polynomials. *Int. J. Syst. Sci.*, 13(2):209–217, Feb. 1982.

[114] C. Hwang and Y. P. Shih. Solution of population balance equations via block pulse functions. *The Chem. Engg. J.*, 25:39–45, 1982.

[115] C. Hwang and Y. P. Shih. Model reduction via Laguerre polynomial technique. *J. Dynamic Syst. Measur. & Cont.*, 105(4):301–304, Dec. 1983.

[116] C. Hwang and Y. P. Shih. Solution of stiff differential equations via generalized block pulse functions. *The Chem. Engg. J.*, 27:81–86, 1983.

[117] C. Hwang and Y. P. Shih. On the operational matrices of block pulse functions. *Int. J. Syst. Sci.*, 17(10):1489–1498, Oct. 1986.

[118] C. Hwang and K. K. Shyu. Analysis and identification of discrete-time systems via discrete Laguerre functions. *Int. J. Syst. Sci.*, 18(10):1815–1824, Oct. 1987.

[119] C. Hwang and K. K. Shyu. Analysis and identification of discrete-time systems via discrete Legendre orthogonal polynomials. *Int. J. Syst. Sci.*, 18(8):1411–1423, Aug. 1987.

[120] C. Hwang and K. K. Shyu. Series expansion approach to the analysis and identification of discrete Hammerstein systems. *Int. J. Cont.*, 47(6):1961–1972, June 1988.

[121] R. Y. Hwang and Y. P. Shih. Model reduction of discrete systems via discrete Chebyshev polynomials. *Int. J. Syst. Sci.*, 15(3):301–308, Mar. 1984.

[122] R. Y. Hwang and Y. P. Shih. Parameter identification of discrete systems via discrete Legendre polynomials. *Comp. & Elect. Engg.*, 12(3/4):155–160, March/Apr. 1986.

[123] Y. Itoh, S. Fukuda, and A. Nii. A microcomputer controlled DC motor speed regulator with load torque estimation. In *Proc. of IEE Conf. on Power Electronics and variable speed drives*, London, 1984.

[124] Y. G. Jan and K. M. Wong. Bilinear system identification by block pulse functions. *J. Franklin Inst.*, 312(5):349–359, Nov. 1981.

[125] Y. G. Jaw and F. C. Kung. Identification of single variable linear time-varying system via block pulse functions. *Int. J. Syst. Sci.*, 15(8):885–893, Aug. 1984.

[126] Z. H. Jiang. Use of block pulse functions for output sensitivity analysis of linear systems. *Int. J. Cont.*, 44(2):407–417, Aug. 1986.

[127] Z. H. Jiang. Block pulse function approach to the identification of MIMO systems and time-delay systems. *Int. J. Syst. Sci.*, 18(9):1711–1720, Sept. 1987.

[128] Z. H. Jiang. New approximation method for inverse Laplace transforms using block pulse functions. *Int. J. Syst. Sci.*, 18(10):1873–1888, Oct. 1987.

[129] Z. H. Jiang and W. Schaufelberger. A new algorithm for single input single output system identification via block pulse functions. *Int. J. Syst. Sci.*, 16(12):1559–1571, Dec. 1985.

[130] Z. H. Jiang and W. Schaufelberger. Recursive formula for the multiple integral using block pulse functions. *Int. J. Cont.*, 41(1):271–279, Jan. 1985.

[131] J. Kalat and P. N. Paraskevopoulos. Solution of multipoint boundary value problems via block pulse functions. *J. Franklin Inst.*, 324(1):73–81, Apr. 1987.

[132] V. R. Karanam, P. A. Frick, and R. R. Mohler. Bilinear system identification by Walsh functions. *IEEE Trans. AC*, AC-23(4):709–713, Aug. 1978.

[133] S. Kawaji. Block pulse series analysis of linear systems incorporating observers. *Int. J. Cont.*, 37(5):1113–1120, May 1983.

[134] S. Kawaji and T. Shiotsuki. Model reduction by Walsh function techniques. *Math. & Comp. Simul.*, 27(5/6):479–484, Oct. 1985.

[135] G. T. Kekkeris. On the analysis of singular systems using orthogonal functions. *IEE Proc., Part - D, CTA*, 133(6):315–316, Nov. 1986.

[136] G. T. Kekkeris. Chebyshev series approach to linear systems sensitivity analysis. *J. Franklin Inst.*, 323(3):273–283, Mar. 1987.

[137] G. T. Kekkeris and P. N. Paraskevopoulos. Analysis of multivariable systems via shifted Chebyshev polynomials. In *Proc. Telekon'84 Conf.*, Helkidiki, Greece, 1984.

[138] M. Korenberg, S. A. Billings, Y. P. Liu, and P. J. McIlroy. Orthogonal parameter estimation algorithm for nonlinear stochastic systems. *Int. J. Cont.*, 48(1):193–210, July 1988.

[139] R. Korte and H. Rake. An optimized method of identifying the parameters of continuous-time systems by discrete-time algorithms. In *Proc. 8th IFAC/IFORS Symp. on Ident. & System Param. Estim.*, Beijing, P. R. China, Aug. 1988.

[140] B. Kouvaritakis and R. G. Cameron. The use of Walsh functions in multivariable limit cycle prediction. *Automatica*, 19(5):513–522, Sept. 1983.

[141] G. Kreisselmeier. An approach to stable indirect adaptive control. *Automatica*, 21(4):425–431, July 1985.

[142] F. C. Kung and H. Lee. Solution of linear state space equations and parameter estimation in feedback systems using Laguerre polynomial expansion. *J. Franklin Inst.*, 314(6):393–403, Dec. 1982.

[143] F. C. Kung and H. Lee. Solution and parameter estimation in linear time-invariant delayed systems using Laguerre polynomial expansion. *J. Dynamic Syst. Measur. & Cont.*, 105(4):297–301, Dec. 1983.

[144] F. C. Kung and D. H. Shih. Analysis and identification of Hammerstein model nonlinear delay systems using block pulse function expansion. *Int. J. Cont.*, 43(1):139–147, Jan. 1986.

[145] C. P. Kwong and C. F. Chen. Linear feedback system identification via block pulse functions. *Int. J. Syst. Sci.*, 12(5):635–642, May 1981.

[146] L. Lee and F. C. Kung. Shifted Legendre series solution and parameter estimation of linear delayed systems. *Int. J. Syst. Sci.*, 16(10):1249–1256, Oct. 1985.

[147] T. T. Lee and Y. F. Chang. Analysis, parameter estimation and optimal control of nonlinear systems via general orthogonal polynomials. *Int. J. Cont.*, 44(4):1089–1102, Oct. 1986.

[148] T. T. Lee and Y. F. Chang. Analysis of time-varying delay systems via general orthogonal polynomials. *Int. J. Cont.*, 45(1):169–181, Jan. 1987.

[149] T. T. Lee and S. C. Tsay. Analysis of linear time-varying systems and bilinear systems via shifted Chebyshev polynomials of the second kind. *Int. J. Syst. Sci.*, 17(12):1757–1766, Dec. 1986.

[150] T. T. Lee and S. C. Tsay. Approximate solutions for linear time-delay systems via the Pade approximation and orthogonal polynomial expansions. *CTAT*, 3(2):111–128, June 1987.

[151] T. T. Lee and Y. F. Tsay. Analysis and optimal control of discrete linear time-varying systems via discrete general orthogonal polynomials. *Int. J. Cont.*, 44(5):1427–1436, Nov. 1986.

[152] W. Leonhard. Microcomputer control of high dynamic performance ac drives – A survey. *Automatica*, 22(1):1–19, Jan. 1986.

[153] F. L. Lewis and B. G. Mertzios. Analysis of singular systems using orthogonal functions. *IEEE Trans. AC*, AC–32(6):527–530, June 1987. Ω

[154] C. T. Liou and Y. S. Chou. Operational matrices of piecewise linear polynomial functions with application to linear time-varying systems. *Int. J. Syst. Sci.*, 18(10):1931–1942, Oct. 1987.

[155] C. T. Liou and Y. S. Chou. Piecewise linear polynomial functions and application to analysis and parameter identification. *Int. J. Syst. Sci.*, 18(10):1919–1929, Oct. 1987.

[156] C. C. Liu and Y. P. Shih. Analysis and optimal control of time-varying systems via Chebyshev polynomials. *Int. J. Cont.*, 38(5):1003–1012, Nov. 1983.

[157] C. C. Liu and Y. P. Shih. Analysis and parameter estimation of bilinear systems via Chebyshev polynomials. *J. Franklin Inst.*, 317(6):373–382, June 1984.

[158] C. C. Liu and Y. P. Shih. Analysis and parameter identification of linear systems via Chebyshev polynomials of second kind. *Int. J. Syst. Sci.*, 16(6):753–759, June 1985.

[159] C. C. Liu and Y. P. Shih. System analysis, parameter estimation and optimal regulator design of linear systems via Jacobi series. *Int. J. Cont.*, 42(1):211–224, July 1985.

[160] C. C. Liu and Y. P. Shih. Model reduction via Chebyshev polynomials. *Comp. & Elect. Engg.*, 12(3/4):89–100, March/Apr. 1986.

[161] Y. Liu and B. D. O. Anderson. Model reduction with time-delay. *IEE Proc., Part – D, CTA*, 134(6):349–367, Nov. 1987.

[162] L. Ljung. *System Identification – Theory for the User*. Prentice Hall, 1987.

[163] L. Ljung and T. Söderström. *Theory and Practice of Recursive Identification*. MIT Press, 1983.

[164] M. Maqusi. Walsh analysis of power-law systems. *IEEE Trans. IT*, pages 144–146, Jan. 1977.

[165] M. Maqusi. On the Walsh analysis of nonlinear systems. *IEEE Trans. EMC*, EMC-20:519–523, Nov. 1978.

[166] W. Marszalek. Analysis of bilinear systems with Picard's method and block pulse operational matrices. *J. Franklin Inst.*, 320(3/4):105–109, Sep./Oct. 1985.

[167] W. Marszalek. On the nature of block pulse operational matrices : Some further results. *Int. J. Syst. Sci.*, 16(6):727–743, June 1985.

[168] B. G. Mertzios. Solution and identification of discrete state space equations via Walsh functions. *J. Franklin Inst.*, 318(6):383–391, Dec. 1984.

[169] B. G. Mertzios, F. L. Lewis, and G. Vachtsevanos. Analysis of singular systems using orthogonal functions. *IEE Proc., Part – D, CTA*, 135(4):323–325, July 1988.

[170] R. H. Middleton. Indirect continuous-time adaptive control. *Automatica*, 23(6):793–795, Nov. 1987.

[171] R. H. Middleton, G. C. Goodwin, D. J. Hill, and D. Q. Mayne. Design issues in adaptive control. *IEEE Trans. AC*, AC–33(1):50–58, Jan. 1988.

[172] B. M. Mohan and K. B. Datta. Delay operational matrix approach for the analysis of time-delay systems. In *Prep. of IMACS/IFAC Int. Symp. on Modelling and Simulation for Control of Lumped and Distributed Parameter Systems*, Lille, France, June 1986.

[173] B. M. Mohan and K. B. Datta. Analysis of time-delay systems via shifted Chebyshev polynomials of first and second kinds. *Int. J. Syst. Sci.*, 19(9):1843–1851, Sept. 1988.

[174] B. M. Mohan and K. B. Datta. Lumped and distributed parameter system identification via shifted Legendre polynomials. *J. Dynamic Syst. Measur. & Cont.*, 110(4):436–440, Dec. 1988.

[175] B. M. Mohan and K. B. Datta. Lumped and distributed parameter system identification via Fourier series. *IEEE Trans. CAS*, 36(11):1454–1458, Nov. 1989.

[176] A. S. Morse. Global stability of parameter adaptive control systems. *IEEE Trans. AC*, AC–25(3):433–439, June 1980.

[177] T. H. Moulden and M. A. Scott. Walsh spectral analysis for ordinary differential equations: Part I – Initial value problems. *IEEE Trans. CAS*, CAS-35(6):742–745, June 1988.

[178] S. G. Mouroutsos and P. N. Paraskevopoulos. Identification of time-varying linear systems using orthogonal functions. *J. Franklin Inst.*, 320(5):249–258, Nov. 1985.

[179] S. G. Mouroutsos and P. D. Sparis. Taylor series approach to system identification, analysis and optimal control. *J. Franklin Inst.*, 319(3):359–371, Mar. 1985.

[180] S. G. Mouroutsos and P. D. Sparis. Parameter identification of a class of time-varying linear systems via polynomial series. *Int. J. Syst. Sci.*, 17(7):969–981, July 1986.

[181] S. G. Mouroutsos and P. D. Sparis. Shift and product Fourier matrices and linear delay-differential equations. *Int. J. Syst. Sci.*, 17(9):1335–1348, Sept. 1986.

[182] S. Mukhopadhyay, A. Patra, and G. P. Rao. Irreducible model estimation for MIMO systems. *Int. J. Cont.*, 53(1):223–253, 1991.

[183] S. Mukhopadhyay and G. P. Rao. An integral equation approach to joint state and parameter estimation in continuous-time MIMO systems. *IEE Proc., Part – D, CTA*, 138, 1991.

[184] K. S. Narendra, Y. H. Lin, and L. S. Valavani. Stable adaptive controller design, Part II : Proof of stability. *IEEE Trans. AC*, AC–25(3):440–448, June 1980.

[185] M. Nieniewski and R. S. Marleau. Digital simulation of an SCR-driven DC motor. *IEEE Trans. IA*, IA-14(4):341–346, July/Aug. 1978.

[186] M. Nieniewski and R. S. Marleau. Mathematical modelling of a digital current control loop for electrical drives. *IEEE Trans. IE*, IE-34(1):107–114, Feb. 1987.

[187] M. Ohkita. An application of rationalized Haar functions to the solution of delay-differential systems. *Math. & Comp. Simul.*, 29(6):477–491, Dec. 1987.

[188] K. R. Palanisamy. *Analysis, optimization and identification of lumped continuous systems with delays via Walsh functions.* PhD thesis, I.I.T. Kharagpur, Kharagpur – 721 302, India, Aug. 1980.

[189] K. R. Palanisamy. Analysis and optimal control of linear systems via single term Walsh series approach. *Int. J. Syst. Sci.*, 12(4):443–454, Apr. 1981.

[190] K. R. Palanisamy. Analysis of nonlinear systems via single term Walsh series approach. *Int. J. Syst. Sci.*, 13(8):929–935, Aug. 1982.

[191] K. R. Palanisamy. A note on the block pulse function operational matrix for integration. *Int. J. Syst. Sci.*, 14(11):1287–1290, Nov. 1983.

[192] K. R. Palanisamy and D. K. Bhattacharya. System identification via block pulse functions. *Int. J. Syst. Sci.*, 12(5):643–647, May 1981.

[193] P. N. Paraskevopoulos. Transfer function matrix identification via Walsh functions. In *Proc., MECO'78*, volume 1, pages 194–197, Athens, Greece, 1978.

[194] P. N. Paraskevopoulos. Chebyshev series approach to system identification, analysis and optimal control. *J. Franklin Inst.*, 316(2):135–157, Aug. 1983.

[195] P. N. Paraskevopoulos. Analysis of singular systems using orthogonal functions. *IEE Proc., Part – D, CTA*, 131(1):37–38, Jan. 1984.

[196] P. N. Paraskevopoulos. Legendre series approach to identification and analysis of linear systems. *IEEE Trans. AC*, AC–30(6):585–589, June 1985.

[197] P. N. Paraskevopoulos. System analysis and synthesis via orthogonal polynomial series and Fourier series. *Math. & Comp. Simul.*, 27(5/6):453–469, Oct. 1985.

[198] P. N. Paraskevopoulos and G. T. Kekkeris. Hermite series approach to system identification, analysis and optimal control. In *Proc., 6th Int. Symp. MECO'83*, volume 1, pages 146–149, Athens, Greece, 1983.

[199] P. N. Paraskevopoulos and R. E. King. Parametric identification using discrete Laguerre polynomials. In *Proc. 4th IFAC Symp. on Ident. & System Param. Estim.*, pages 21–27, Tbilisi, USSR, Sept. 1976.

[200] P. N. Paraskevopoulos, P. D. Sparis, and S. G. Mouroutsos. The Fourier series operational matrix for integration. *Int. J. Syst. Sci.*, 16(2):171–176, Feb. 1985.

[201] A. Patra, P. V. Bhaskar, and G. P. Rao. A package for simulation and parameter estimation of continuous-time dynamical systems. In *Proc. 8th IFAC/IFORS Symp. on Ident. & System Param. Estim.*, pages 1959–1963, Beijing, P. R. China, Aug. 1988.

[202] A. Patra and G. P. Rao. Some recursive parameter estimation schemes for continuous-time models – An assessment. In *Proc. of the 12th National Systems Conf.*, pages 343–348, Coimbatore, India, Dec. 1988.

[203] A. Patra and G. P. Rao. Continuous-time approach to self-tuning control: Algorithm, implementation and assessment. *IEE Proc., Part – D, CTA*, 136(6):333–340, Nov. 1989.

[204] A. Patra and G. P. Rao. General hybrid orthogonal functions and some potential applications in systems and control. *IEE Proc., Part – D, CTA*, 136(4):157–163, July 1989.

[205] A. Patra and G. P. Rao. General hybrid orthogonal functions - A new tool for the analysis of power electronic systems. *IEEE Trans. IE*, IE-36(3):413–424, Aug. 1989.

[206] A. Patra and G. P. Rao. Parameter estimation in a converter-driven d.c. motor system via general hybrid orthogonal functions. *IEEE Trans. IE*, IE-38(4):398–403, Oct. 1991.

[207] A. E. Pearson and F. C. Lee. Efficient parameter identification for a class of bilinear differential systems. In *Proc. 7th IFAC Symp. on Ident. & System Param. Estim.*, pages 161–165, University of York, U. K., July 1985. Pergamon Press.

[208] A. E. Pearson and F. C. Lee. On the identification of polynomial input-output differential systems. *IEEE Trans. AC*, AC-30(8):778–782, Aug. 1985.

[209] A. E. Pearson and F. C. Lee. Parametric identification of discontinuous nonlinearities. In *Proc. 7th IFAC Symp. on Ident. & System Param. Estim.*, University of York, U. K., July 1985. Pergamon Press.

[210] A. E. Pearson and C. Y. Wuu. Some computational aspects in identifying differential delay systems. In *Proc. 1983 ACC*, pages 1255–1256, San Francisco, June 1983.

[211] A. E. Pearson and C. Y. Wuu. System identification of pure delay with finite time data. In *Proc. 19th IEEE CDC*, Albuquerque, New Mexico, Dec. 1983.

[212] A. E. Pearson and C. Y. Wuu. Decoupled delay estimation in the identification of differential delay systems. *Automatica*, 20(6):761–772, Nov. 1984.

[213] R. Pintelon, K. Schoukens, and J. Renneborg. Estimation of continuous-time system parameters in s- or z-domain: A comparison. In *Proc. 8th IFAC/IFORS Symp. on Ident. & System Param. Estim.*, Beijing, P. R. China, Aug. 1988.

[214] S. Puthenpura and N. K. Sinha. Robust bootstrap method for joint estimation of states and parameters of a linear system. *J. Dynamic Syst. Measur. & Cont.*, 108(3):255–263, Sept. 1986.

[215] V. Ranganathan, A. N. Jha, and V. S. Rajamani. Recursive parameter estimation algorithms for bilinear and nonlinear systems using a Laguerre polynomial approach. *Int. J. Cont.*, 44(2):419–426, Aug. 1986.

[216] G. P. Rao. *Piecewise Constant Orthogonal Functions and Their Application to Systems and Control*, volume 55 of *LNCIS*. Springer Verlag, Berlin, 1983.

[217] G. P. Rao and K. R. Palanisamy. A new operational matrix for delay via Walsh functions and some aspects of its algebra and applications. In *Proc. 5th National Systems Conf.*, pages 60–61, PAU, Ludhiana, Punjab, India, Sept. 1978.

[218] G. P. Rao and K. R. Palanisamy. Walsh stretch matrices and functional differential equations. *IEEE Trans. AC*, AC–27(1):272–276, Feb. 1982.

[219] G. P. Rao and K. R. Palanisamy. Improved algorithms for parameter identification in continuous systems via Walsh functions. *IEE Proc., Part – D, CTA*, 130(1):9–16, Jan. 1983.

[220] G. P. Rao and K. R. Palanisamy. Analysis of time-delay systems via Walsh functions. *Int. J. Syst. Sci.*, 15(1):9–30, Jan. 1984.

[221] G. P. Rao, K. R. Palanisamy, and T. Srinivasan. Extension of computation beyond the limit of initial normal interval in Walsh series

analysis of dynamical systems. *IEEE Trans. AC*, AC–25(2):317–319, Apr. 1980.

[222] G. P. Rao and A. Patra. Continuous-time approaches to combined state and parameter estimation in linear continuous systems. In *Proc. 8th IFAC/IFORS Symp. on Ident. & System Param. Estim.*, pages 1287–1291, Beijing, P. R. China, Aug. 1988.

[223] G. P. Rao, D. C. Saha, T. M. Rao, A. Bhaya, and K. Aghoramurthy. A microprocessor-based system for on-line parameter identification in continuous dynamical systems. *IEEE Trans. IE*, IE-29(3):197–201, Aug. 1982.

[224] G. P. Rao and L. Sivakumar. System identification via Walsh functions. *Proc. IEE*, 122(10):1160–1161, Oct. 1975.

[225] G. P. Rao and L. Sivakumar. Identification of time-lag systems via Walsh functions. *IEEE Trans. AC*, AC–24(5):806–808, Oct. 1979.

[226] G. P. Rao and L. Sivakumar. Transfer function matrix identification in MIMO systems via Walsh functions. *Proc. IEEE*, 69(4):465–466, Apr. 1981.

[227] G. P. Rao and L. Sivakumar. Order and parameter identification in continuous linear systems via Walsh functions. *Proc. IEEE*, 70(7):764–766, July 1982.

[228] G. P. Rao and L. Sivakumar. Piecewise linear system identification via Walsh functions. *Int. J. Syst. Sci.*, 13(5):525–530, May 1982.

[229] G. P. Rao and T. Srinivasan. Analysis and synthesis of dynamic systems containing time-delays via block pulse functions. *Proc. IEE*, 125(9):1064–1068, Oct. 1978.

[230] G. P. Rao and T. Srinivasan. Solution of certain nonlinear functional differential equations via block pulse functions. In *Proc. 5th National Systems Conf.*, pages 287–290, PAU, Ludhiana, Punjab, India, Sept. 1978.

[231] G. P. Rao and T. Srinivasan. An optimal method of solving differential equations characterizing the dynamics of a current collection system for an electric locomotive. *J. Inst. Math. & Appln.*, 25(4):329–342, June 1980.

[232] G. P. Rao and S. G. Tzafestas. Analysis, control and identification of distributed parameter and time-delay systems using piecewise constant orthogonal basis functions – an overview. In *Proc. 24th IEEE CDC*, Dec. 1985.

[233] G. P. Rao and S. G. Tzafestas. A decade of piecewise constant orthogonal functions in systems and control. *Math. & Comp. Simul.*, 27(5/6):389–407, Oct. 1985.

[234] C. E. Rohrs, L. Valavani, M. Athans, and G. Stein. Robustness of continuous-time adaptive control algorithms in the presence of unmodelled dynamics. *IEEE Trans. AC*, AC–30(9):881–889, Sept. 1985.

[235] V. A. Romanov and V. A. Semeran. Algorithm for identifying the dynamic characteristics of objects by means of orthogonal Walsh functions. *Autom. and Remote Cont.*, 34:601–607, 1973.

[236] S. Sagara and Z. Y. Zhao. An improved algorithm for on-line identification in continuous systems via numerical integration. In *Proc. SICE'87*, pages 1077–1080, Hiroshima, Japan, July 1987.

[237] S. Sagara and Z. Y. Zhao. On-line identification of continuous systems using linear integral filter. In *Proc. of 19th JAACE*, pages 41–46, Fukuoka, Japan, Oct. 1987.

[238] S. Sagara and Z. Y. Zhao. Parameter identification in continuous systems via numerical integration. Technical Report 4, Kyushu University, Kyushu, Japan, Aug. 1987.

[239] S. Sagara and Z. Y. Zhao. Numerical integration approach to on-line parameter identification of continuous systems in presence of measurement noise. In *Proc. 8th IFAC/IFORS Symp. on Ident. & System Param. Estim.*, Beijing, P. R. China, Aug. 1988.

[240] D. C. Saha, B. B. P. Rao, and G. P. Rao. Structure and parameter identification in linear continuous lumped systems – The Poisson moment functional approach. *Int. J. Cont.*, 36(3):477–491, Sept. 1982.

[241] D. C. Saha and G. P. Rao. Identification of lumped linear systems in the presence of unknown initial conditions via Poisson moment functionals. *Int. J. Cont.*, 31(4):637–644, Apr. 1980.

[242] D. C. Saha and G. P. Rao. Identification of lumped linear time-varying parameter systems via Poisson moment functionals. *Int. J. Cont.*, 32(4):709–721, Oct. 1980.

[243] D. C. Saha and G. P. Rao. Identification of lumped linear continuous systems – The Poisson moment functional approach. In *Proc. of Int. Conf. on Systems Theory and Applications*, pages A163–A173, PAU, Ludhiana, India, Dec. 1981.

[244] D. C. Saha and G. P. Rao. Identification of lumped linear systems in the presence of small unknown time-delays – The Poisson moment functional approach. *Int. J. Cont.*, 33(5):945–951, May 1981.

[245] D. C. Saha and G. P. Rao. A general algorithm for parameter identification in lumped continuous systems – The Poisson moment functional approach. *IEEE Trans. AC*, AC–27(1):223–225, Feb. 1982.

[246] D. C. Saha and G. P. Rao. Transfer function matrix identification in MIMO systems via Poisson moment functionals. *Int. J. Cont.*, 35(4):727–738, Apr. 1982.

[247] D. C. Saha and G. P. Rao. *Identification of Continuous Dynamical Systems – The Poisson Moment Functional (PMF) Approach*, volume 56 of *LNCIS*. Springer Verlag, Berlin, 1983.

[248] P. Sannuti. Analysis and synthesis of dynamic systems via block pulse functions. *Proc. IEE*, 124(6):569–571, June 1977.

[249] T. K. Sarkar, N. Radhakrishna, and H. Chen. Survey of various z-domain to s-domain transformations. *IEEE Trans. IM*, IM–35(4):508–520, Dec. 1986.

[250] L. A. Shieh and R. E. Yates. Solving inverse Laplace transform, linear and nonlinear state equations using block pulse functions. *Comp. & Elect. Engg.*, 6:3–17, 1979.

[251] L. S. Shieh, C. K. Yeung, and B. C. McInnis. Solution of state space equations via block pulse functions. *Int. J. Cont.*, 28(3):383–392, Sept. 1978.

[252] D. H. Shih and F. C. Kung. The shifted Legendre approach to nonlinear system analysis and identification. *Int. J. Cont.*, 42(6):1399–1410, Dec. 1985.

[253] D. H. Shih and F. C. Kung. Analysis and parameter estimation of a scaled system via shifted Legendre polynomials. *Int. J. Syst. Sci.*, 17(3):401–408, Mar. 1986.

[254] D. H. Shih and F. C. Kung. Analysis and parameter estimation of nonlinear systems via shifted Chebyshev expansions. *Int. J. Syst. Sci.*, 17(2):231–240, Feb. 1986.

[255] D. H. Shih and F. C. Kung. Shifted Legendre approach to the analysis and identification of a linear delayed system with a nonlinear gain. *IEE Proc., Part – D, CTA*, 133(3), May 1986.

[256] D. H. Shih, F. C. Kung, and C. M. Chao. Laguerre series approach to the analysis of a linear control system incorporating observers. *Int. J. Cont.*, 43(1):123–128, Jan. 1986.

[257] Y. M. Shih. Block pulse function analysis of time-varying and nonlinear networks. *J. Chinese Inst. of Engr.*, 1(2):43–52, Feb. 1978.

[258] Y. P. Shih and W. K. Chia. Piecewise constant solutions of integral equations via Walsh functions. *J. Chinese Inst. of Engr.*, 1(1):81–85, Jan. 1978.

[259] Y. P. Shih and C. Hwang. Application of block pulse functions in dynamic simulation. *Comp. & Chem. Engg.*, 6(1):7–13, Jan. 1982.

[260] Y. P. Shih, C. Hwang, and W. K. Chia. Parameter estimation of delay systems via block pulse functions. *J. Dynamic Syst. Measur. & Cont.*, 102(3):159–162, Sept. 1980.

[261] Y. P. Shih and C. C. Liu. Parameter estimation of time-varying systems via Chebyshev polynomials of the second kind. *Int. J. Syst. Sci.*, 17(3):849–858, June 1986.

[262] N. K. Sinha. Estimation of transfer functions of continuous systems from sampled data. *Proc. IEE*, 119(5):612–614, May 1972.

[263] N. K. Sinha and G. J. Lastman. Identification of continuous-time multivariable systems from sampled data. *Int. J. Cont.*, 35(1):117–126, Jan. 1982.

[264] N. K. Sinha and S. Puthenpura. Choice of the sampling interval for the identification of continuous-time systems from samples of input-output data. *IEE Proc., Part – D, CTA*, 132(6):263–267, Nov. 1985.

[265] N. K. Sinha and Q. J. Zhou. Discrete-time approximation of multivariable continuous systems. *IEE Proc., Part – D, CTA*, 130(3):103–110, May 1983.

[266] N. K. Sinha and Q. J. Zhou. State estimation using block pulse functions. *Int. J. Syst. Sci.*, 15(4):341–350, Apr. 1984.

[267] L. Sivakumar. *Some aspects of system identification leading to Walsh function approach*. PhD thesis, I.I.T. Kharagpur, Kharagpur – 721 302, India, Sept. 1978.

[268] A. Y. Sivaramakrishnan and M. C. Srisailam. Walsh series solution of second order state space systems. *Int. J. Syst. Sci.*, 16(5):633–639, May 1985.

[269] O. J. M. Smith. Estimator for dynamic systems using operating records. *J. Dynamic Syst. Measur. & Cont.*, 109(3):253–267, Sept. 1987.

[270] P. D. Sparis. Application of the operational matrix of differentiation for the identification of time-varying linear systems using polynomial series. *IEE Proc., Part – D, CTA*, 134(3):180–186, May 1987.

[271] P. D. Sparis and S. G. Mouroutsos. A comparative study of the operational matrices for integration and differentiation for orthogonal polynomial series. *Int. J. Cont.*, 42(3):621–638, Sept. 1985.

[272] P. D. Sparis and S. G. Mouroutsos. The operational matrix for polynomial series transformation. *Int. J. Syst. Sci.*, 16(9):1173–1184, Sept. 1985.

[273] P. D. Sparis and S. G. Mouroutsos. The operational matrix of differentiation for orthogonal polynomial series. *Int. J. Cont.*, 44(1):1–15, July 1986.

[274] T. Srinivasan. *Analysis of dynamical systems via block pulse functions*. PhD thesis, I.I.T. Kharagpur, Kharagpur – 721 302, India, 1979.

[275] Z. Trazaska. Computation of the block pulse solution of singular systems. *IEE Proc., Part – D, CTA*, 133(4):191–192, July 1986.

[276] M. J. Tsai, C. K. Chen, and F. C. Kung. Analysis of linear time-varying systems by shifted Legendre polynomials. *J. Franklin Inst.*, 318(4):275–282, Oct. 1984.

[277] S. C. Tsay and T. T. Lee. Solution of integral equations via Taylor series. *Int. J. Cont.*, 44(3):701–709, Sept. 1986.

[278] S. C. Tsay and T. T. Lee. Analysis and optimal control of linear time-varying systems via general orthogonal polynomials. *Int. J. Syst. Sci.*, 18(8):1579–1594, Aug. 1987.

[279] Y. F. Tsay and T. T. Lee. Solution of discrete scaled systems via general discrete orthogonal polynomials. *Int. J. Cont.*, 44(6):1715–1724, Dec. 1986.

[280] S. G. Tzafestas. Walsh series approach to lumped and distributed system identification. *J. Franklin Inst.*, 305(4):199–220, Apr. 1978.

[281] S. G. Tzafestas, J. Anoussis, and C. Papastergiou. Dynamic nuclear reactivity computation using block pulse function expansion. *Int. J. Model. & Simul.*, 4(2):73–76, 1984.

[282] H. Unbehauen and G. P. Rao. *Identification of Continuous Systems*. North Holland, Amsterdam, 1987.

[283] S. Vajda, P. Valkó, and K. R. Godfrey. Direct and indirect least squares methods in continuous-time parameter estimation. *Automatica*, 23(6):707–718, 1987.

[284] T. L. Van, L. D. C. Tam, and N. V. Houtte. On direct algebraic solutions of linear differential equations using Walsh transforms. *IEEE Trans. CAS*, CAS-22(5):419–422, May 1975.

[285] J. Vlassenbroeck and R. van Dooren. A Chebyshev technique for solving nonlinear optimal control problems. *IEEE Trans. AC*, AC-33(4):333–340, Apr. 1988.

[286] W. M. Walmsley and W. A. Evans. Application of the fast Walsh transform to system identification. *J. D'electronique sur Technologies de Pointe pour le Traitement des Signaux*, pages 307–318, 1975.

[287] C. H. Wang and R. S. Marleau. System identification via generalized block pulse operational matrices. *Int. J. Syst. Sci.*, 16(11):1425–1430, Oct. 1985.

[288] C. H. Wang and R. S. Marleau. Application of generalized block pulse operational matrices for the approximation of continuous-time systems. *Int. J. Syst. Sci.*, 17(9):1269–1278, Sept. 1986.

[289] C. H. Wang and R. S. Marleau. Recursive computational algorithms for the generalized block pulse operational matrix. *Int. J. Cont.*, 45(1):195–201, Jan. 1987.

[290] C. H. Wang and Y. P. Shih. Explicit solutions of integral equations via block pulse functions. *Int. J. Syst. Sci.*, 13(7):773–782, July 1982.

[291] M. L. Wang and R. Y. Chang. Solution of linear dynamic systems with initial or boundary value conditions by shifted Legendre approximations. *Int. J. Syst. Sci.*, 14(3):343–353, Mar. 1983.

[292] M. L. Wang, R. Y. Chang, and S. Y. Yang. Analysis and optimal control of time-varying systems via generalized orthogonal polynomials. *Int. J. Cont.*, 44(4):895–910, Oct. 1986.

[293] M. L. Wang, R. Y. Chang, and S. Y. Yang. Identification of a single-variable linear time-varying system via generalized orthogonal polynomials. *Int. J. Syst. Sci.*, 18(9):1659–1671, Sept. 1987.

[294] M. L. Wang, K. S. Chen, and C. K. Chou. Solutions of integral equations via modified Laguerre polynomials. *Int. J. Syst. Sci.*, 15(6):661–672, June 1984.

[295] M. L. Wang, Y. J. Jan, and R. Y. Chang. Analysis and parameter identification of time-delay linear systems via generalized orthogonal polynomials. *Int. J. Syst. Sci.*, 18(9):1645–1658, Sept. 1987.

[296] M. L. Wang, S. Y. Yang, and R. Y. Chang. Analysis of systems with multiple time-varying delays via generalized block pulse functions. *Int. J. Syst. Sci.*, 18(3):543–552, Mar. 1987.

[297] M. L. Wang, S. Y. Yang, and R. Y. Chang. Application of generalized block pulse functions to a scaled system. *Int. J. Syst. Sci.*, 18(8):1495–1503, Aug. 1987.

[298] M. L. Wang, S. Y. Yang, and R. Y. Chang. Application of generalized orthogonal polynomials to parameter estimation of time-invariant and bilinear systems. *J. Dynamic Syst. Measur. & Cont.*, 109(1):7–13, Mar. 1987.

[299] M. L. Wang, S. Y. Yang, and R. Y. Chang. New approach for parameter identification via generalized orthogonal polynomials. *Int. J. Syst. Sci.*, 18(3):569–579, Mar. 1987.

[300] S. Y. Wang. The application of block pulse functions in identification of parameters in nonlinear time-varying systems. In *Proc. 6th IFAC Symp. on Ident. & System Param. Estim.*, pages 919–924, 1982.

[301] S. Y. Wang. The design of suboptimal inputs for identifying parameters in linear time-varying system. *IEEE Trans. AC*, AC–29(7):633–637, July 1984.

[302] A. H. Whitfield and N. Messali. Continuous system order identification from input-output data. *Int. J. Cont.*, 46(4):1399–1410, Oct. 1987.

[303] A. H. Whitfield and N. Messali. Integral equation approach to system identification. *Int. J. Cont.*, 45(4):1431–1445, Apr. 1987.

[304] B. W. Williams. Complete state space digital computer simulation of chopper fed DC motors. *IEEE Trans. IECI*, IECI-25(3):255–260, Aug. 1978.

[305] B. Wittenmark and K. J. Åström. Practical issues in the implementation of self-tuning control. *Automatica*, 20(5):595–606, Sept. 1985.

[306] P. C. Young. The determination of the parameters of a dynamic process. *Radio and Electronic Engineer*, 29:345–361, 1965.

[307] P. C. Young. Process parameter estimation and self-adaptive control. In P. J. Hammond, editor, *Theory of Self-adaptive Systems*. Plenum Press, New York, 1966.

[308] P. C. Young. Parameter estimation for continuous-time models – A survey. *Automatica*, 17(1):23–39, Jan. 1981.

[309] S. Zaman and A. N. Jha. Parameter identification of nonlinear systems using Laguerre operational matrices. *Int. J. Syst. Sci.*, 16(5):625–631, May 1985.

[310] C. Zervos, P. R. Belanger, and G. A. Dumont. On PID controller tuning using orthonormal series identification. *Automatica*, 24(2):165–175, Mar. 1988.

[311] Z. F. Zhao and M. B. Zarrop. Sequential test signal generation for parameter estimation in continuous-time systems. Technical Report 623, Control Systems Center, UMIST, Manchester, England, Nov. 1984.

[312] O. C. Zienkiewicz. *The Finite Element Method.* Tata McGraw Hill, New Delhi, 1979.

Index

actuator saturation 81
approximation of a discontinu-
     ous function 7

block pulse functions (BPF) 1,
     44, 76, 97

chopper driven d.c. motor 34
     duty cycle 38–40
     free-wheeling interval 37
     steady-state waveforms 39
     transient response 40
completeness 6
continuous basis functions (CBF)
     1
continuous-time system identifi-
     cation 47
controllability 89
converter-driven d.c. motor 32,
     50
     parameter estimation 58, 68
     steady-state waveforms 35
     transient response 36

digital filtering 48
discrete-time (DT) 52
distributed parameter systems 88

emulator 74
extension of solution of state equa-
     tion 18

finite element (FE) methods 21
Fourier series 1

fully controlled bridge converter
     2
function expansion 6
functional differential equation 97,
     98

Galerkin method 23
general hybrid orthogonal func-
     tions (GHOF) 3, 48
     definition 3
     multidimensional 89
     properties 5
generalized Fourier coefficients 6
generalized least squares 51, 58,
     60, 62, 63
generalized orthogonal polynomi-
     als (GOP) 12
GHOF spectral analysis of dy-
     namical systems 11–19
GHOF spectrum 13

Haar functions (HF) 1
harmonic balance 41

identification of CT systems 47–
     71
initial interval 18
inner product 5

Kronecker product 16

least squares 14
Legendre polynomials 4, 7, 33,
     38, 43

limit cycle 41, 44
line voltage 31
load disturbance 80
lumped linear system 13

minimum-time control 89
Monte-Carlo simulation 50
MRAC 73
MSMT formula 18
MSST formula 18

numerical analysis of dynamical
        systems 21
numerical integration 48

observability 89
OF-based solutions 2
operational matrices 13, 88
optimal control 88
orthogonal functions (OF) 1, 48,
        89
orthogonal polynomials 1
orthogonality 5

parameter convergence 81, 82
parameter estimation 55
        converter driven d.c. motor
        55
Parseval's condition 6
partial calculus 89
piecewise constant basis functions
        (PCBF) 1
Poisson moment functionals 48
power electronics 2
predictive control 74

recursive
        computation of multiple in-
        tegrals 52
        least squares (LS) 54
relative order 48
Runge-Kutta methods 30

sampling 47, 48
saw-tooth waveform 7
SCR controlled d.c. drive 13
self-tuning control (STC) 47, 73
        CT algorithm 76–80
        CT approaches 73–86
        DT algorithm 81–86
        hybrid 73
        implementation 79
        implicit/ explicit/ direct/ in-
        direct schemes 74–75
simulation of SCR controlled drives
        30–40
sliding mode control 89
solution of state equation 16
square root algorithm 80
SSMT formula 19
SSST formula 20
stability 89
suboptimal control 89

Taylor series 48
time-delay 48, 88, 88
Tustin transformation 81

UD factorization 80

Van der Pol's oscillator 41, 87
        half-period 44
        solution 43–45
variational method 23, 24

Walsh functions (WF) 1, 30
weighting function 5, 22

Lecture Notes in Control and Information Sciences

Edited by M. Thoma

1992–1996 Published Titles:

Vol. 167: Rao, Ming
Integrated System for Intelligent Control
133 pp. 1992 [3-540-54913-7]

Vol. 168: Dorato, Peter; Fortuna, Luigi;
Muscato, Giovanni
Robust Control for Unstructured
Perturbations: An Introduction
118 pp. 1992 [3-540-54920-X]

Vol. 169: Kuntzevich, Vsevolod M.; Lychak,
Michael
Guaranteed Estimates, Adaptation and
Robustness in Control Systems
209 pp. 1992 [3-540-54925-0]

Vol. 170: Skowronski, Janislaw M.;
Flashner, Henryk; Guttalu, Ramesh S. (Eds)
Mechanics and Control. Proceedings of the
4th Workshop on Control Mechanics,
January 21-23, 1991, University of
Southern California, USA
302 pp. 1992 [3-540-54954-4]

Vol. 171: Stefanidis, P.; Paplinski, A.P.;
Gibbard, M.J.
Numerical Operations with Polynomial
Matrices: Application to Multi-Variable
Dynamic Compensator Design
206 pp. 1992 [3-540-54992-7]

Vol. 172: Tolle, H.; Ersü, E.
Neurocontrol: Learning Control Systems
Inspired by Neuronal Architectures and
Human Problem Solving Strategies
220 pp. 1992 [3-540-55057-7]

Vol. 173: Krabs, W.
On Moment Theory and Controllability of
Non-Dimensional Vibrating Systems and
Heating Processes
174 pp. 1992 [3-540-55102-6]

Vol. 174: Beulens, A.J. (Ed.)
Optimization-Based Computer-Aided
Modelling and Design. Proceedings of the
First Working Conference of the New IFIP
TC 7.6 Working Group, The Hague, The
Netherlands, 1991
268 pp. 1992 [3-540-55135-2]

Vol. 175: Rogers, E.T.A.; Owens, D.H.
Stability Analysis for Linear Repetitive
Processes
197 pp. 1992 [3-540-55264-2]

Vol. 176: Rozovskii, B.L.; Sowers, R.B.
(Eds)
Stochastic Partial Differential Equations and
their Applications. Proceedings of IFIP WG
7.1 International Conference, June 6-8,
1991, University of North Carolina at
Charlotte, USA
251 pp. 1992 [3-540-55292-8]

Vol. 177: Karatzas, I.; Ocone, D. (Eds)
Applied Stochastic Analysis. Proceedings of
a US-French Workshop, Rutgers University,
New Brunswick, N.J., April 29-May 2,
1991
317 pp. 1992 [3-540-55296-0]

Vol. 178: Zolésio, J.P. (Ed.)
Boundary Control and Boundary Variation.
Proceedings of IFIP WG 7.2 Conference,
Sophia-Antipolis, France, October 15-17,
1990
392 pp. 1992 [3-540-55351-7]

Vol. 179: Jiang, Z.H.; Schaufelberger, W.
Block Pulse Functions and Their
Applications in Control Systems
237 pp. 1992 [3-540-55369-X]

Vol. 180: Kall, P. (Ed.)
System Modelling and Optimization.
Proceedings of the 15th IFIP Conference,
Zurich, Switzerland, September 2-6, 1991
969 pp. 1992 [3-540-55577-3]

Vol. 181: Drane, C.R.
Positioning Systems - A Unified Approach
168 pp. 1992 [3-540-55850-0]

Vol. 182: Hagenauer, J. (Ed.)
Advanced Methods for Satellite and Deep
Space Communications. Proceedings of
an International Seminar Organized by
Deutsche Forschungsanstalt für Luft-und
Raumfahrt (DLR), Bonn, Germany,
September 1992
196 pp. 1992 [3-540-55851-9]

Vol. 183: Hosoe, S. (Ed.)
Robust Control. Proceesings of a Workshop
held in Tokyo, Japan, June 23-24, 1991
225 pp. 1992 [3-540-55961-2]

Vol. 184: Duncan, T.E.; Pasik-Duncan, B.
(Eds)
Stochastic Theory and Adaptive Control.
Proceedings of a Workshop held in
Lawrence, Kansas, September 26-28,
1991
500 pp. 1992 [3-540-55962-0]

Vol. 185: Curtain, R.F. (Ed.); Bensoussan,
A.; Lions, J.L.(Honorary Eds)
Analysis and Optimization of Systems:
State and Frequency Domain Approaches
for Infinite-Dimensional Systems.
Proceedings of the 10th International
Conference, Sophia-Antipolis, France, June
9-12, 1992.
648 pp. 1993 [3-540-56155-2]

Vol. 186: Sreenath, N.
Systems Representation of Global Climate
Change Models. Foundation for a Systems
Science Approach.
288 pp. 1993 [3-540-19824-5]

Vol. 187: Morecki, A.; Bianchi, G.;
Jaworeck, K. (Eds)
RoManSy 9: Proceedings of the Ninth
CISM-IFToMM Symposium on Theory and
Practice of Robots and Manipulators.
476 pp. 1993 [3-540-19834-2]

Vol. 188: Naidu, D. Subbaram
Aeroassisted Orbital Transfer: Guidance
and Control Strategies
192 pp. 1993 [3-540-19819-9]

Vol. 189: Ilchmann, A.
Non-Identifier-Based High-Gain Adaptive
Control
220 pp. 1993 [3-540-19845-8]

Vol. 190: Chatila, R.; Hirzinger, G. (Eds)
Experimental Robotics II: The 2nd
International Symposium, Toulouse,
France, June 25-27 1991
580 pp. 1993 [3-540-19851-2]

Vol. 191: Blondel, V.
Simultaneous Stabilization of Linear
Systems
212 pp. 1993 [3-540-19862-8]

Vol. 192: Smith, R.S.; Dahleh, M. (Eds)
The Modeling of Uncertainty in Control
Systems
412 pp. 1993 [3-540-19870-9]

Vol. 193: Zinober, A.S.I. (Ed.)
Variable Structure and Lyapunov Control
428 pp. 1993 [3-540-19869-5]

Vol. 194: Cao, Xi-Ren
Realization Probabilities: The Dynamics of
Queuing Systems
336 pp. 1993 [3-540-19872-5]

Vol. 195: Liu, D.; Michel, A.N.
Dynamical Systems with Saturation
Nonlinearities: Analysis and Design
212 pp. 1994 [3-540-19888-1]

Vol. 196: Battilotti, S.
Noninteracting Control with Stability for
Nonlinear Systems
196 pp. 1994 [3-540-19891-1]

Vol. 197: Henry, J.; Yvon, J.P. (Eds)
System Modelling and Optimization
975 pp approx. 1994 [3-540-19893-8]

Vol. 198: Winter, H.; Nüßer, H.-G. (Eds)
Advanced Technologies for Air Traffic Flow
Management
225 pp approx. 1994 [3-540-19895-4]

Vol. 199: Cohen, G.; Quadrat, J.-P. (Eds)
11th International Conference on
Analysis and Optimization of Systems –
Discrete Event Systems: Sophia-Antipolis,
June 15–16–17, 1994
648 pp. 1994 [3-540-19896-2]

Vol. 200: Yoshikawa, T.; Miyazaki, F. (Eds)
Experimental Robotics III: The 3rd
International Symposium, Kyoto, Japan,
October 28-30, 1993
624 pp. 1994 [3-540-19905-5]

Vol. 201: Kogan, J.
Robust Stability and Convexity
192 pp. 1994 [3-540-19919-5]

Vol. 202: Francis, B.A.; Tannenbaum, A.R.
(Eds)
Feedback Control, Nonlinear Systems,
and Complexity
288 pp. 1995 [3-540-19943-8]

Vol. 203: Popkov, Y.S.
Macrosystems Theory and its Applications:
Equilibrium Models
344 pp. 1995 [3-540-19955-1]

Vol. 204: Takahashi, S.; Takahara, Y.
Logical Approach to Systems Theory
192 pp. 1995 [3-540-19956-X]

Vol. 205: Kotta, U.
Inversion Method in the Discrete-time
Nonlinear Control Systems Synthesis
Problems
168 pp. 1995 [3-540-19966-7]

Vol. 206: Aganovic, Z.;.Gajic, Z.
Linear Optimal Control of Bilinear Systems
with Applications to Singular Perturbations
and Weak Coupling
133 pp. 1995 [3-540-19976-4]

Vol. 207: Gabasov, R.; Kirillova, F.M.;
Prischepova, S.V.
Optimal Feedback Control
224 pp. 1995 [3-540-19991-8]

Vol. 208: Khalil, H.K.; Chow, J.H.;
Ioannou, P.A. (Eds)
Proceedings of Workshop on Advances in
Control and its Applications
300 pp. 1995 [3-540-19993-4]

Vol. 209: Foias, C.; Özbay, H.;
Tannenbaum, A.
Robust Control of Infinite Dimensional
Systems: Frequency Domain Methods
230 pp. 1995 [3-540-19994-2]

Vol. 210: De Wilde, P.
Neural Network Models: An Analysis
164 pp. 1996 [3-540-19995-0]

Vol. 211: Gawronski, W.
Balanced Control of Flexible Structures
280 pp. 1996 [3-540-76017-2]

Vol. 212: Sanchez, A.
Formal Specification and Synthesis of
Procedural Controllers for Process Systems
248 pp. 1996 [3-540-76021-0]